Ford Farm Toys

David Reed

Schiffer Publishing Ltd
4880 Lower Valley Road, Atglen, PA 19310 USA

Dedication

To Patsy, Lindsey and the Cat

Disclaimer

The text and products pictured in this book are from the collection of the author, his publisher, or various private collectors. This book is not sponsored, endorsed or otherwise affiliated with any of the companies whose products are represented herein. They include: Ford Motor Company, Henry Ford and Son, Inc, Fordson, Ertl, Arcade, Whitehead and Kales, A.C. Williams, Bing, Kenton, Bates Steele Mule, Gilson Riecke, Tekno Bogg, Tootsietoy, Jordan Product, Lewis Galoob Toys, Micro Machines, Pioneer of Power, Ezra Brooks, Heritage China Company, Shirley Nance, Arthur Hanes, Ralston Toy Novelty Company, Ralstoy, Danbury Mint, Britains, Corgi, Crescent, Lesney, Matchbox, Fun Ho, New Holland, Ford New Holland, Doe Company, Bob Gray, *Toy Tractor Times*, Ford Ferguson, Hubley, Product Miniature, Aluminum Model Toys, Slik Toy, Modern Toys, Marusan, Dearborn, Georgia Marble, Lowell Davis, Foxfire, Popular Imports, Spec Cast, Carter Tru-Scale, Franklin Mint, Lansing Manufacturing Company, Irwin Toy Corporation, Woods Brothers, McDonalds, Mattel Toy Company (Barbie), Filmways Corporation, Central Broadcasting System, Nostalig, County of England, Dave Sharp, Majorette, Galanite, Gabriel, Olti (Strike), Maisto, *Toy Farmer*, Wheatland, Minneapolis Moline, Lonestar, *Replica*, Processed Plastics, Polistil, Styrofoam, New Ray, Haybine, Jouef, National Farm Toy Museum, Siku, Buddy L, Pacesetter, Plasto, Cragstan, John Deere, Versatile, Ford Versatile, Sperry New Holland, Hover, Mack Truck, First Gear, Fiat, New Holland, Geotech, New Holland Geotech, Scale Models, Western Auto, and Firestone, among others. This book is derived from the author's independent research.

Most of the items and products in this book may be covered by various copyrights, trademarks, and logotypes. Their use herein is for identification purposes only. All rights are reserved by their respective owners.

Copyright © 2003 by David Reed
Library of Congress Control Number: 2002113630

All rights reserved. No part of this work may be reproduced or used in any form or by any means—graphic, electronic, or mechanical, including photocopying or information storage and retrieval systems—without written permission from the copyright holder.

"Schiffer," "Schiffer Publishing Ltd. & Design," and the "Design of pen and ink well" are registered trademarks of Schiffer Publishing Ltd.

Designed by John P. Cheek
Cover Design by Bruce M. Waters
Type set in Americana XBd BT / Souvenir Lt BT

ISBN: 0-7643-1744-X
Printed in China
1 2 3 4

Published by Schiffer Publishing Ltd.
4880 Lower Valley Road
Atglen, PA 19310
Phone: (610) 593-1777; Fax: (610) 593-2002
E-mail: Schifferbk@aol.com
Please visit our web site catalog at
www.schifferbooks.com
We are always looking for people to write books on new and related subjects. If you have an idea for a book, please contact us at the above address.

This book may be purchased from the publisher.
Include $3.95 for shipping.
Please try your bookstore first.
You may write for a free catalog.

In Europe, Schiffer books are distributed by
Bushwood Books
6 Marksbury Avenue
Kew Gardens
Surrey TW9 4JF England
Phone: 44 (0) 20 8392 8585
Fax: 44 (0) 20 8392 9876
E-mail: Bushwd@aol.com
Free postage in the UK. Europe: air mail at cost.

Contents

Acknowledgments 4
Introduction 5
Chapter One: Pre-Fordson and Fordson 1907–1938 6
Chapter Two: Ford "N" Series 1939–1952 23
Chapter Three: Ford Jubilee Series 1953–1954 34
Chapter Four: Ford 600 and 700 Series 37
Chapter Five: Ford 900 Series 41
Chapter Six: Ford 1000 and 2000 Series 49
Chapter Seven: Ford 4000 and 5000 Series 52
Chapter Eight: Ford 6000 and 7000 Series 63
Chapter Nine: Ford 8000 and 9000 Series 73
Chapter Ten: Ford TW Series 85
Chapter Eleven: Ford Power Star Series 94
Chapter Twelve: Ford Industrial Series 99
Chapter Thirteen: Ford Articulated Four-Wheel Drive 109
Chapter Fourteen: Ford Lawn and Garden 112
Chapter Fifteen: Ford Machinery 114
Chapter Sixteen: Ford Farm Sets 124
Chapter Seventeen: Ford Dealer Trucks 130
Bibliography 143
Index 144

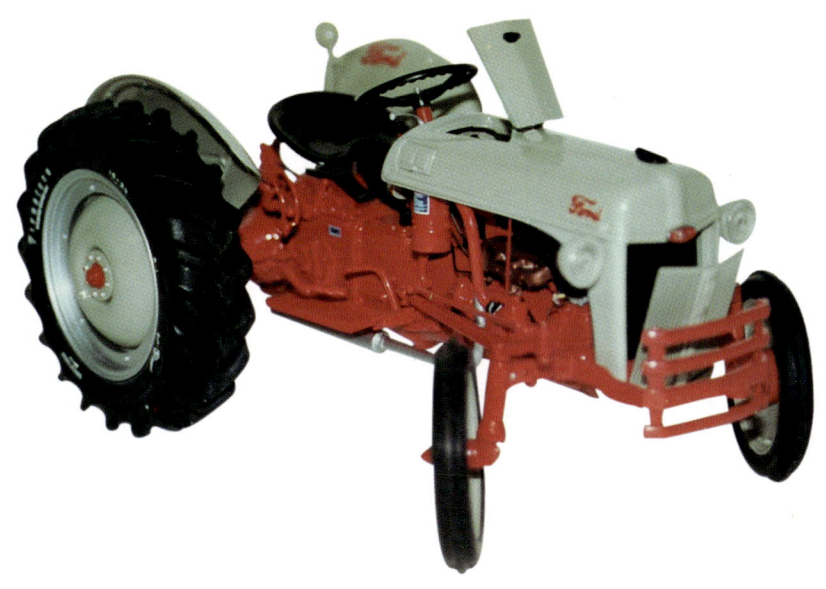

Acknowledgments

I would like to thank Jim Newman, John Hill and Steve Drake for allowing items from their collections to be photographed and included as part of this book. We've had a lot of good times exchanging information and stories about toys in our collections.

Many thanks to my sister Sue Latham and our good friend Mary McCleery for the use of their paintings as backgrounds in the photographs.

A thank you to Jerry Conner for his help in proofreading the text and finding misspelled words.

Also thanks to Sharon Snordgrass and Earl Friedmeyer for the use of photo supplies.

To our editors Jeff Synder and Donna Baker, our thanks for your patience and kindness.

Finally, a special thank you to my wife Patsy and my daughter Lindsey for their support, especially their typing and computer skills.

Introduction

To all those who have been collecting Ford toy tractors for many years and for all new collectors, I would like to welcome you to my first edition of *Ford Farm Toys*.

The Ford Tractor Company has gone through many changes in the last few years, making Ford farm toys more desirable. In the early 1990s, Fiat of Italy began the process of purchasing Ford New Holland from the Ford Motor Company. The Ford oval found on the front of the tractor was gone by 1995 and the Ford name seen on the side of the tractors was gone by the year 2000. The Ford name was dropped from these toys and the New Holland name is now seen in its place. Because this is a book on Ford farm toys, only those toys bearing the name of Ford and Ford New Holland are mentioned.

The prices quoted in this book reflect auction prices as well as the purchase price of items paid by myself and other Ford collectors. Mint condition, in the box original items are always the most sought after by collectors and command premium prices. With the removal of the Ford name, older Ford farm toys have increased in value and will continue to do so with the passage of time.

Everyone has a reason for their interest in collecting toys, and most interests reflect toys from their childhood.

There have been numerous companies that have produced Ford farm toys in the United States and worldwide. The older farm toys produced by Hubley are the ones that I enjoy the most because these are toys I remember from my own childhood.

I hope you enjoy reading this book that features my pictures and descriptions of Ford farm toys and related items from my personal collection as well the collections of others.

Price Disclaimer

The "Value" in this book reflects the price of a toy in mint condition without the box. New in the box toys (N.I.B.) or new in the package (N.I.P.) toys reflect the price of a toy in mint condition in its original shipping carton or box. A toy in mint condition will usually double in price if its box is in exceptionally nice condition, sometimes even more if the toy is considered rare.

The price of toys varies according to condition, region, and rarity of the toy. The author cannot be held responsible for any monetary losses that an individual might incur as a result of using this price guide.

Chapter One
Pre-Fordson and Fordson 1907-1938

1907 Auto Plow™: Tractors were high on Henry Ford's list of priorities in the early years of the Ford Motor Company. As a youth, Ford walked many a weary mile behind a slow moving mule and plow and was determined to make farming what it ought to be – the most pleasant and profitable profession in the world. Henry Ford began making plans for the production of farm tractors as early as 1905. This first experimental auto plow, as he called it, was completed in 1907 (Williams, 1985). This 1/16 scale replica of Ford's auto plow is a one of a kind model made by the author from brass and spare toy tractor parts. This model took three months to complete, costing a total of thirteen dollars. Value: $50, N.I.B n/a.

1915-1916 Henry Ford and Son: Henry Ford's company directors failed to show enthusiasm for his tractor project in 1915. Ford then set up a completely separate company to research and develop his tractor. The new company was named Henry Ford and Son Incorporated. This 1/16 scale model was painted flat black and given a round emblem, which reads, "Henry Ford and Son." This was added to the grill of the Ertl open-casting Fordson. This model was made to resemble Ford's pre-production Fordson. Value $20, N.I.B. n/a.

1928 Arcade™ Fordson: This photograph shows a 6 inch Arcade Fordson Model "F" tractor, which appears to be on its way to Latham, a small Missouri town ten miles west of this old covered bridge. At 2 1/2 miles per hour, this little smooth-wheeled Arcade toy tractor should arrive at Grover's market in about four hours. *Background of photograph courtesy of Sue Latham*. Value: $125, N.I.B. $350.

Arcade "F" Fordson and Sulky Hay Rake: This little Fordson tractor and sulky rake still carry their original paint. The tractor measures 5.75 inches long with a red body and lime green wheels. Unlike most Fordson toys, the crank and the name were never included in this model. The driver was cast into the main body halves. A close examination of this toy reveals the Arcade name cast into the inside of the engine casting. Even though the Arcade sulky rake isn't a Ford, it was made in this smaller size to accompany the little Fordson toys of its day. The rake was red like the tractor but had nickel-plated wheels. Value Tractor $125, N.I.B. $250; Value Rake $125, N.I.B. n/a.

Fordson "F" with "WK" Wheels: This 6 inch long toy Arcade Fordson was manufactured around 1932 and originally had a nickel-plated driver. These toys are believed to have been used by salesmen for a company by the name of Whitehead and Kales, a manufacturer of the solid rubber disc wheels as seen on this toy. A full-sized tractor with this type of wheel was often used at railroad depots to pull mail and freight wagons. Highway builders and other industrial contractors used this type of wheeled tractor because it was easy on concrete and other types of hard surfaces and the owner didn't have to worry about flat tires. Value: $175, N.I.B. $300.

Fordson Model "F": A.C. Williams of the United States made this Fordson model "F" in a 3.75 inch model. It could be purchased between 1940 and 1950. Value $125, N.I.B. $250.

Tin Wind-Up Fordson: Bing of Germany made this Fordson tractor and could be purchased in the early 1900s with a trailer or wagon. Value $125, N.I.B. $250.

Road Roller and Kenton™ Grader: Information on this 4.6 inch long Fordson is non-existent in my private library. Cast iron Fordsons are fairly easy to find and had several different manufacturers between the 1920s and 1950s. This toy, however, sports its own original red and gray paint with a wood front roller assembly. Connected to the roller is a 7.5 inch long Kenton grader in nice original paint. Although this toy isn't a Fordson, it probably spent many hours behind one, possibly owned by a happy child in 1930. Fordson Tractor. Value $100, N.I.B. $200; Kenton Grader. Value $150, N.I.B. n/a.

Fordson Model "F" with Bates Steel Mule™: This little Fordson tractor was a very popular tractor that almost every farmer could afford. Its popularity tempted companies by the hundreds, to manufacture accessories for the Fordson. One such accessory manufactured by Bates Machine and Tractor Company of Joliet Illinois, made a steel crawler conversion, which greatly increased traction. These were used not only by farmers but also by road and construction contractors. This Bates conversion added $295 dollars to the original cost of the tractor, which in some years, sold for less than $300.

Gilson Riecke introduced a nice model of the Fordson with Bates Steel Mule in approximately 1/20 scale. The undercarriage was highly detailed with each individual track section being pinned to the next. Since each of these models was hand made, there was a waiting list of one and one half years before my Fordson could be built. Mr. Riecke also made this Fordson in a companion model, which had a casting of the driver in it as well. I paid $275 dollars for this model in 1989. Value $300, N.I.B. n/a.

Fordson Model F: This Fordson was made in the 1/43 scale by Tenko Bogg of Sweden. This toy could be purchased in the late 1990s. Value $50, N.I.B. $100.

1/87 scale Fordsons: Among the smallest Fordsons in my collection, are these two little gray model "F" Fordsons sitting on top of this Tootsietoy transporter. The tractor models were available in plastic model kits that took a magnifying glass, tweezers, and a lot of patience to assemble. The tractor on the left is the industrial version, which came with dual disc rear wheels. The tractor on the right represents the more common farm version with red spoke wheels. Both kits were made by Jordan Products of Canton, Ohio and sold for $1.50 each. I do not know how long these kits were available; however, I purchased the set in 1984, several years after they were first produced.

The little yellow industrial Fordson in the foreground was made by Lewis Galoob Toys, better known as Micro Machines™ of southern San Francisco, California. This Fordson toy came in a set of five farm vehicles and could be purchased at discount stores in 1989 for about four dollars. Some collectors bought these tractors in large quantities hoping to make a profit. At toy shows the asking price was ten dollars and many collectors refused to pay this price leaving the sellers in the lurch. An easily found red and white version was introduced in 1985. Only those who wanted the toys for their collection were buying at this time. Value: $5, N.I.B. $10.

Reproduction Fordson with Loader: This reproduction Fordson was manufactured in the 1/16 scale. It was available in 2000 for twenty- five dollars. Value $10, N.I.B. $25.

Ezra Brooks Whisky Bottle: This 9 inch long Fordson manufactured by Ezra Brooks of Frankfort, Kentucky is not to be considered a child's toy, at least not while it contained its .08 quart, 90 proof, Kentucky spinning whisky. But it is definitely a collector's item. The sculptor, Shirley Nance of Heritage China Company, designed this as a salute to the Tri-State Gasoline Engine and Tractor Association. This was the sixth annual showing of their collection at the Jay County Indiana Fair Grounds held in Portland, Indiana in 1971. Value $25, N.I.B. $45.

Kansas Toy and Novelty™ Fordson: In 1923, Arthur Hanes, an automobile mechanic from Clifton, Kansas began molding toys from his garage and sold them to local stores in the Kansas area. Most of his models were automobiles, however, some were farm-related items. One such item, as shown in this photograph, resembled a Fordson Tractor. These tractors could easily be spotted by their unusual wheel pattern. In later years, Ralston Toy and Novelty Company of Ralston, Nebraska, acquired some of the toy molds and began producing a variation of the toy with round spoke-type front wheels. It is possible, however, that Ralston could have made toy Fordsons in either wheel variation. Value $50, N.I.B. n/a.

Ertl Closed Casting Model "F" Fordson: In 1917 Henry Ford was ready to put his first tractor into production. The problem was, a tractor company from Minnesota was already using the Ford surname of one of their own employees. Mr. Ford decided to use a nickname of the words, Henry Ford and Son, which was simply Fordson on his first production tractor.

Probably the most common toy Fordson is this 1/16 scale model "F" Fordson made by the Ertl company of Dyersville, Iowa from 1969 until 1989. In the twenty years this toy was produced by Ertl, several different box variations could be found with only two main casting variations. They are commonly referred to as the rare open engine casting and the more common closed engine casting as shown in the photograph. Value $20, N.I.B. $35.

Green Acres™ Fordson: This early version of the Ertl Fordson shows the open engine casting. These were only made in the first years of production and are difficult to find today in good, original condition. The toy shown in this photograph was made in the late 1960s and came in the Green Acres box. It was a promotional toy for the Filmways TV show known as Green Acres, shown on CBS in the 1960s. Stars on the show were Eddie Albert as Oliver Douglas and Eva Gabor as Lisa Douglas. Occasionally, Oliver could be seen on the television show farming with an early Fordson tractor. Value $10, N.I.B. $95.

Canadian Special Edition™ Fordson: This special edition of the 1/16 scale English style Fordson tractor was produced by Ertl to commemorate the forth annual Canadian International Farm Equipment Show held in Toronto, Ontario on February sixth through the ninth, 1990.

The tractor was blue with orange wheels and featured a Canadian maple leaf decal with the inscription "C.I.F.E.S. 1990 Special Edition," cast into the side. *Photograph courtesy of Jim Newman.* Value $24, N.I.B. $50.

1923 English Fordson: This Ertl Fordson toy has the early open engine casting and was mildly customized by painting the steel wheels Ford Blue. This paint combination was commonly seen on the 1923 English Fordson. Value $20, N.I.B. n/a.

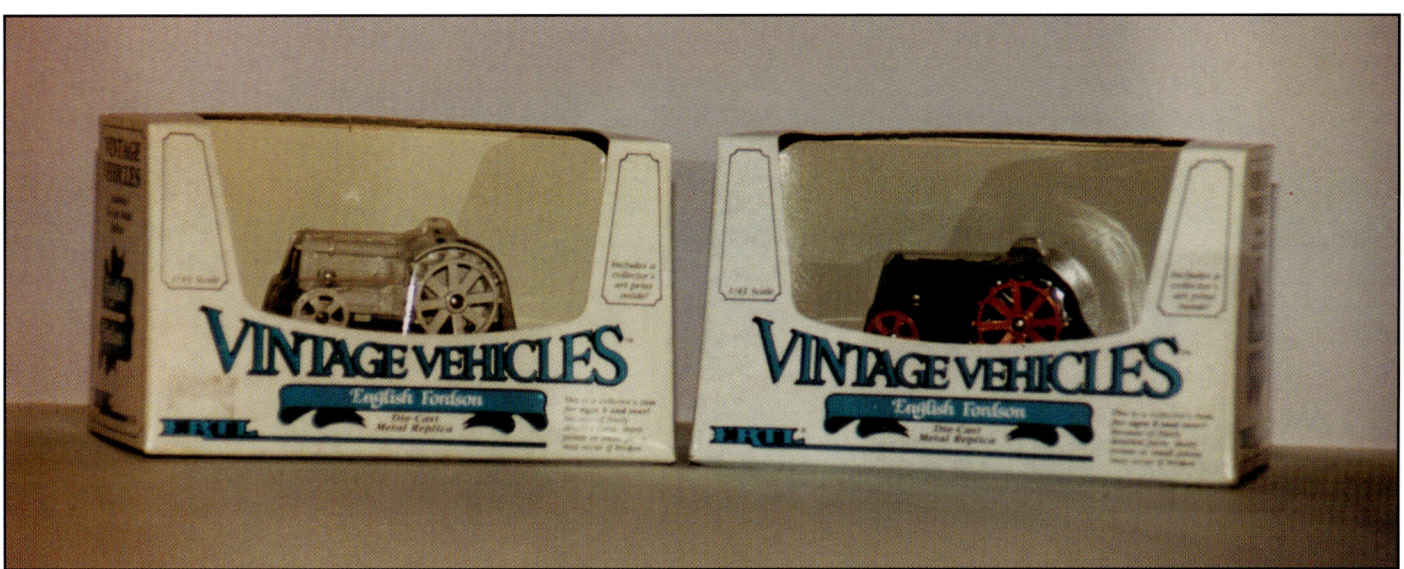

The 1/43 scale English Fordson: In late 1980s, Ertl began a new "Vintage Vehicles™" series of toys made in the 1/43 scale. These toys were made in China and could be purchased in the United States for five dollars. Two Fordson toys could be found in this series with the first one resembling the all-gray version commonly seen in the United States. It was available to collectors in this area in late 1987. The blue Fordson with the orange wheels is a replica of the tractors commonly seen in England. This toy model was available at area stores beginning in April 1990. Value $5 each, N.I.B. $15 each.

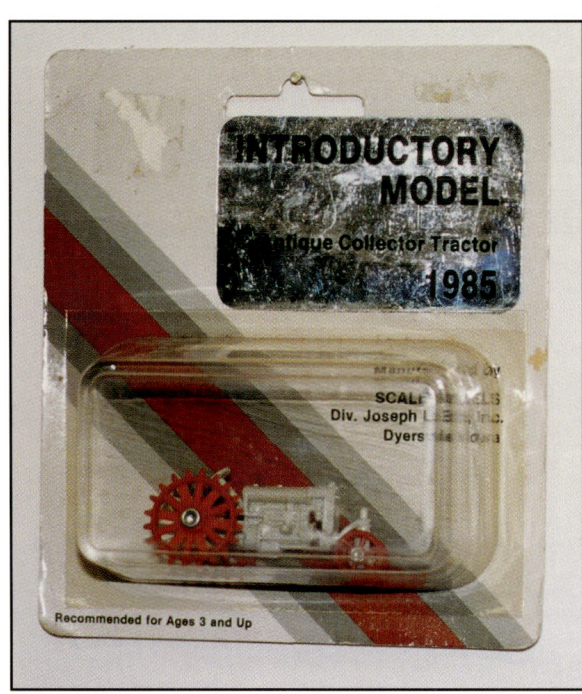

1936 All Around™ with Steel Wheels: Scale Models made this little plastic 1/64 scale Fordson in 1985. It came in a special introductory package for one dollar and fifty cents. Value $3, N.I.P. $5.

1936 All Around on Rubber: This 1/64 scale die cast tricycle Fordson was manufactured by Scale Models It could be purchased for one dollar and fifty cents in 1990. Value $3, N.I.P $5.

1930 Fordson N's: In late 1985, Scale Models manufactured this little blue with orange steel wheel English Fordson. They were produced in a limited edition of 5000 units. Each serial numbered toy was made in the popular 1/16 scale and could be purchased for approximately thirty dollars. A decal on the dash read "Ford Motor Company LTD. England."

The orange Fordson on the right, started its life as a duplicate to the Fordson on the left. The blue paint was changed to orange, rubber tires with spoke wheels were added, and a silver exhaust was installed. This gave the 1930 model the appearance of an actual 1939 Ford built for the British government in the event that the war on mainland Europe spread to England. Three thousand full-sized tractors were ordered and kept serviced and ready by Fordson dealers in England. These tractors were used on farms in England to produce food during World War II. Value $35, N.I.B. $75.

New 1/16 Model "F" Fordson: A welcomed toy to a Ford toy collector was the totally new casting by Ertl of the "F" series Fordson. This more detailed version had several improvements including full crown fenders, automobile type steering and my favorite, an engine crank.

The first tractor in the photograph shows the third variation, which came without the rear fenders. It did have a dash decal that read, "Fordson, Ford Motor Company, Detroit, Michigan, U.S.A." This variation first appeared in toy stores in early 1993 and was priced around sixteen dollars. Value $16, N.I.B. $30.

The second tractor in the photograph shows the regular edition, sometimes referred to as the "Sandbox Edition." It has the full fenders, no engine, crank, or dash decal. It *was available at toy outlets in the mid 1990s for a twenty*-dollar bill. Value $10, N.I.B. $20.

The last toy Fordson in the photograph also came out in the mid 1990s and was found at the local Ford dealerships. With a price tag of forty-five dollars, it would hardly be classified as a toy. Value $25, N.I.B. $50.

Danbury Mint™ Model "F" Fordson: This Fordson was made in 1/16 scale by Danbury Mint of Norwal**k**, Connecticut. It could be purchased in 2000 for $127. Background in the photograph courtesy of Mary McCleery. Value $100, N.I.B. $127.

1/64 scale Fordsons: These four Fordsons were more of a toy than a collector's item. Very few collectors added them to their collections therefore in years to come they may be a seldom seen Fordson item. They were made by the Ertl Company and were commonly sold in the $2 price range. The first toy was made in 1982 when the Smurf™ was featured in a popular cartoon series. This is an orange tractor with a blue Smurf driver. The words, "Ol McSmurf," were found on each fender. The next, a blue Fordson was made in 1985 for the "Toy Farmer"™ as a collectible and featured Zeke as the driver. Later in 1987 Zeke was found on a red Fordson with rubber rear tires.

The card to the right shows a green Fordson with a driver everyone knows, Porky Pig™ It came out in a series of six different Warner Brothers characters in different vehicles. This tractor was made in 1989 and had a production run of only 2500. Value $3 each. N.I.B. $5 each.

Lillyput™ E27N: Barely as big as a quarter, is a Britains's E27N English Fordson. This particular tractor was made around 1953 and came in the Lillyput Series. This tractor was very popular with model railroaders. Value $75, N.I.B. $125.

Britains Fordson Major™: This tractor was made by Britains of England in 1956 in the 1/32 scale, and could be purchased for around four dollars. Value $75, N.I.B $150.

Fordson E27N: Britains manufactured this tractor in the 1/32 scale. It could be purchased from 1948 to 1957. Value $75, N.I.B. $150.

E27N with Henry Ford: Britains also manufactured this 1/32 Fordson model "F" and placed it on a plaque. It could be purchased in 1999 for seventy-five dollars. Value $25, N.I.B. $75.

Fordson Half Track™ Tractor: This English farmer seems proud of his Fordson Super Major™ with half-tracks. Corgi of Wales made this tractor in the 1/43 scale from 1962 until 1964. This type of modification isn't commonly seen on farm tractors here in the United States. Value $75, N.I.B. $200.

Crescent Dexta and Wagon: In 1957, Crescent of Great Britain made this Fordson Dexta Farm Set in 1/25 scale. It consisted of an orange Fordson tractor and a red four wheel wagon. A blue variation with a red wagon was also available. Value $50, N.I.B. $100.

Fordson Power Major™: This Fordson Power Major was made in the 1/43rd scale by Corgi of England. It could be purchased from 1964 through 1966 for about four dollars. *Background in photograph courtesy of Mary McCleery.* Value $50, N.I.B. $100.

Lesney Fordson: In 1963, Lesney of England made this tractor trailer set in the 1/42 scale. It was a Fordson Super Major with orange wheels and was found in the Matchbox King Size Series™ as toy K-11. Value $15, N.I.B. $30.

Fun Ho Fordson Major: Fun Ho of New Zealand made this Fordson Major in the 1/87 scale. It could be purchased in 1967 for one dollar. Value $10, N.I.B. $25.

1/32 scale Fordson: This photograph shows a pair of 1/32 scale Fordson tractors that were produced by Ertl for the Ford 1991 Parts Mart Show. The blue tractor on the right resembles a Ford 5000. Both tractors had "Parts Mart Special Edition," on the rear fenders. Each tractor came in a special edition box and could be purchased for around twelve dollars each. Value $10 each, N.I.B. $25 each.

Fordson 5000: The Fordson 5000 Super Major was made in 1988 in a special limited edition. If featured a beautiful gray and blue paint scheme with cast metal wheel rims on both the front and rear. Other added features were a black and silver vertical exhaust, painted steering wheel, and a collector's insert on the tool box which read, "1988 Special Edition Diesel 5000." This model was available from Ford New Holland dealers for thirty-five dollars. Value $25, N.I.B. $50.

Fordson Super Major: In 1988 Ertl introduced a new tractor which, of course, meant a new casting. It was the English version of the Ford tractor, better known as the Fordson Super Major. It is made to 1/16 scale and had a blue body with plastic reddish-orange wheel rims. It also had the vertical exhaust with a black muffler and carried a price tag of seventeen dollars. Value $10, N.I.B. $25.

Parts Mart Super Major: The second special edition Super Major in the 1/16 scale made by Ertl was this solid blue Fordson with Ford gray cast metal wheel rims. It had the black painted steering wheel with a special collector's insert cast into the tool box which read, "Parts Mart Special Edition 1990." It came with a gray vertical exhaust and muffler. It could be purchased for thirty-two dollars. Value $20, N.I.B. $35.

Fordson Super Major: This Fordson Super Major was made in the 1/16 scale by Ertl. It could be purchased in 1991 for twenty dollars. Value $10, N.I.B. $20.

Doe Dual Drive™ or Triple D™: In the early 1960's four-wheel drive tractors were not commonly available. The Doe Company of Maldon, England took up the manufacturing rights of a farmer's invention. This invention consisted of taking two Super Major tractors combined together to form one articulated machine.

Steve Drake, a toy collector from Tennessee, built replicas of this tractor combination using the Ertl 1/16 scale Super Major toy. *Photographs courtesy of Steve Drake.* Value $100. each N.I.B. n/a.

Ford Historical Set: Ertl first introduced this set in the early 1990s and it could be purchased for nine dollars. Several later tractor models were made by changing the paint scheme and decals, but these tractors were never added to the set. Value $15, N.I.B. $25.

Agbusinessmen Fundraising Tractor: This is a promotional tractor produced by a group of businessmen who were involved in the promotion of agriculture through the Republican Party. It was a fundraiser in the 1992 presidential election. It was a great campaign fundraiser, but President Bush Sr. did not win the election in 1992.

The truck and tractor in the photo features a Ford truck with an elephant riding a Fordson tractor. There were 375 items produced featuring the Ford truck and Fordson tractor combination. When one of these promotional items was purchased, a copy of the signed thank you letter from President Bush Sr. and a political badge was included.

The truck was made by Winross® and the tractor was made by Ertl of Dyersville, Iowa. Value $75, N.I.B. $175.

Chapter Two
Ford "N" Series 1939-1952

Arcade™ 9N: In 1940, Arcade made this 6.5 inch 9N Ford tractor of cast iron. Later variations would be found with stamped steel rear fenders and if you got lucky, you might even have found one of these with a two bottom mounted plow. Value $275, N.I.B. $1500.

Reproduction Cast Iron 2N: The 1939 9N Ford Ferguson on the left closely resembles an original Arcade Ford from late 1940s. It even has the Arcade name in the body molding; however Bob Gray manufactured this toy in 1970. A cousin shown on the right resembles a 1942 2N wartime steel-wheeled model. The manufacturer of this tractor was Pioneer Toy Company. This 1988 reproduction could be purchased for about twenty-five dollars. Value $30, N.I.B. n/a.

Toy Tractor Times™ 9N: Henry Ford and Harry Ferguson introduced the full sized 9N Ford Ferguson tractor at Ford's Fairlane Estates in June 1939. Powered by a little twenty-eight horse power four cylinder engine, the customer received a lot of tractor for $585.

If you missed out on that deal, the *Toy Tractor Times*, a farm toy magazine from Osage Iowa, had a replica produced in the 1/16 scale. Its magazine subscribers could order up to four of these 1/16 scale anniversary models, for a cost of forty-seven dollars and fifty cents each.

The little Toy Tractor Times tractor had a nickel plated hood and grill, special TTT collector insert, horizontal grill decal, and came in a tenth anniversary box. Value $40, N.I.B. $65.

Precision Series™ "9N" Ford: Probably one of the nicest Ford toy models in any Ford collection is the 1/16 scale "Precision Series" made in China by Ertl. This one represents the early version "9N" with the aluminum grill and hood of the 1939 model. The grill appears to be chromed, but on the full-sized early production Ford Ferguson tractors, the hood and grill were cast of aluminum because dye stamping equipment for steel was not available. The aluminum hoods and grills proved to be very fragile and were replaced over the years. The first precision Ford hit the market in time for Christmas of 1994. If Santa didn't bring this toy, you could expect to pay over one hundred dollars. However, for the detail in these models, they were several hundred dollars less than other popular detailed models produced by private individuals. N.I.B. $195.

Ford 9N: Both gray "N" series Fords shown here are 9Ns, however, close examination reveals that the gray paint scheme is almost the only resemblance the toys have to the full sized 9N tractor. The casting for this toy was borrowed from the Ertl 1/16 scale 8N, produced in 1985, so the wheel rims tend to look out of place on this toy Ford. My Ford tractor dealer only received six of the twelve toy 9Ns he ordered, so I felt somewhat fortunate to have been able to pay the thirty dollars this toy was bringing. Unofficial reports in toy magazines stated that Ertl made only 2000 of this model. This might have been why the price tag was three times higher than the ten dollars the regular edition was bringing. Value $40, N.I.B. $50.

The 9N on the right, shown with its 1939 horizontal grill, was also made by Ertl to commemorate the fiftieth anniversary of the famous "Hand Shake" agreement between Henry Ford and Harry Ferguson and the production of the first 9N Ford Ferguson tractor made in 1939 (Pripps, Robert 1991). This special edition tractor and plow cost thirty-five dollars when it was first released in 1989. Value $40, N.I.B. $60.

Ford Plastic 8Ns: These 8Ns were made in the 1/24 scale over a period of forty years. The green 8N to the far left was made by Hubley around 1950 and was produced in a variety of colors and wheel variations. Scale Models made the gray and the red 8N also in a variety of colors. Scale Models would give away one of these tractors when you toured their plant during the National Farm Toy Show. Value $15, N.I.B. n/a.

1/64 scale 9N: These three 9N toys are basically the same except for the packaging. The one to the left was the 1939, 9N in the Ford New Holland Package. It was basically Ertl's earlier 1/64 scale 8N except for the gray paint and the horizontal grill decal. It first appeared in 1994 in the three dollar price range.

The middle 9N was packaged in a nice little display box and came in a set with nine other boxed tractors. The set commemorated Ertl's fiftieth anniversary in 1995. To get this tractor, I had to purchase the complete set, which sold for thirty dollars.

The 9N to the right was manufactured in 1995 for the National Farm Toy Museum as a fundraiser. Value $3 each, N.I.P. $10 each.

Farm Progress Show 9N Ford: This tractor was made by Scale Models in 1996 in the 1/64 scale, and could be purchased for three dollars. Value $5, N.I.P. n/a.

1942 Wartime Ford Ferguson: Echoing through the boardroom of the Ford Tractor Company one could imagine hearing these words, "Keep production up! Keep costs down! Save critical metal with the Ferguson System!" The year? It was in 1942 that Ford Motor Company was forced to drastically change specification of the popular little 9N Ford Ferguson. Due to material shortages caused by World War II, the 1942 model, now designated as the 2N had no battery, starter, generator, electrical system, or rubber tires. It did however, have a crank, magneto, and steel wheels.

Ertl of Dyersville, Iowa, made a surprisingly accurate model of this wartime Ford Ferguson tractor in 1995. Manufactured in the popular 1/16 scale, it was the second Ford in the Precision Series with a price ranging between eighty to one hundred dollars. It probably won't end up in many sandboxes. Value $100, N.I.B. $125.

Early 8N Ford: The tractor in the photo is a plastic 8N manufactured in 1945 by Product Miniature Company. What makes this 8N unusual is the plastic front axel. The axel proved to be fragile, however, and couldn't take bumping into too many chair legs without breaking. Product Miniature soon began casting the tractor with metal front axles. *Photograph courtesy of Jim Newman.* Value $275, N.I.B. $450.

Early Product Miniature Ford 8N: In July 1947, the first new 8N Ford tractor rolled off the assembly line at the Rouge Plant, west of Detroit Michigan. The most noticeable differences between the old 9N and new 8N was the lighter gray paint with bright red casting. Another noticeable difference was the Ford script on the hood and rear fenders. Gone forever was the Ferguson badge that had its home on the front of over 250 thousand Ford-Ferguson tractors.

When two Ford collectors start talking about their collection, the question "Do you have any plastic Fords?" will arise. More than likely they are talking about the 8N, Jubilee, Model 600, and Model 900 all made of plastic in the 1/12 scale.

The photograph shows an early 8N without Ford script on the rear fender. Also, the generator is located on the right hand side of the engine as on the full-sized 8N Fords. This toy Ford has the metal front axle that proved to be much sturdier than the earlier models with the plastic axle. Value $250, N.I.B. $400.

Third Product Miniature 8N Ford: This 1/12 scale 8N Ford was also made by Product Miniature. I call this the third variation, which has over twenty casting changes of the previous model. The most noticeable difference is the location of the generator, which has now been moved to the left side of the engine. This change was made on the full sized tractors to make room for the side mounted distributor. Value $250, N.I.B. $400.

Early Wind up 8N: If you had two dollars and sixty-nine cents, in the early 1950s, you could buy one of these plastic 8N Fords with a wind up motor. Aluminum Model Toys of Detroit, Michigan, manufactured this toy. This unique variation has slick rear tires. Value $275. N.I.B. $450.

Wind Up 8N Ford: It only takes a quick glance to determine a difference in this plastic 8N compared to those in the previous photographs. It is very similar to the Product Miniature toy 8N's, yet it is quite different. The most obvious difference is the addition of the extra body length to accommodate a wind up motor. This motor is mounted neatly inside the transmission and rear axle housing. A crank found on the right hand side would activate the motor. AMT (Aluminum Model Toys Inc.) of Detroit, Michigan made this Ford toy in the early 1950s. It could be purchased for two dollars and sixty-nine cents. Value $225, N.I.B. $375.

Japanese Tin Tractors: You probably won't find this pair of Fords listed in the toy price guide. Information is hard to locate, but I call them Fords because of a few close similarities to the real machines.

The red toy on the left has several characteristics of the 8N. The tractor is 5.25 inches long and made of painted tin. This tractor was sold as a single toy as shown in the photograph or in a farm set, which included a driver and three-piece machinery set. A company called Modern Toys made the set in Japan during the 1950s.

The blue tractor on the right resembles the Jubilee series. It was also made in Japan of painted tin, however, a company known as Marusano manufactured this tractor. Value $15 each, N.I.B. $35 each.

Tootsietoy Toy 8N and Disc Harrow: If one picture is indeed worth a thousand words, the scenery found on the shipping carton of this Tootsietoy 8N Ford tells much more than I could possibly describe. The toy is nice, but hours could be spent studying the farm scenery, which appears on the sides and bottom of this box. By the way, the little white sticker on the front of the box reads, "Handyman Lumber Company 10, 59, 98 cents." Value $50, N.I.B. $125.

Tootsietoy 8N: Tootsietoy of Chicago, Illinois made these nice little 8N Fords in the 1950s. They came in the 1/32 scale and had several castings and wheel variations. The two main variations were a tractor with a red hood and the front-end loader and a tractor with a silver hood that came with a front-end loader or a disc. Either set sold new for about one dollar. Value $50, N.I.B. $125 each.

Tractors of the Past 8N Set: This 8N toy set was made in 1987 by Ertl and is referred to by collectors as the "8N Tractor of the Past." The set included a 1/16 scale 8N red belly tractor, and a 1/43 scale 8N with a front-end loader. The unique characteristic of this set is the little gray loader found in the 1/43 scale tractor. In regular production, the loader was painted red. Many collectors bought several sets, thinking they would be a good investment, but this was not the case. Collectors who originally paid eighteen dollars per set started dumping them at toy shows for around ten dollars. Value $25, set N.I.B. $50 set.

8N Ford with Dearborn Plow: In 1987, Ertl produced this set of Ford 8N's with Dearborn two bottom plows. The one on the left was a collector's edition with chrome decals and stationary plow. Its price tag was in the twenty-eight dollar range. The sandbox edition on the right with the removable Dearborn plow could be found for only eighteen dollars. Value $14 each, N.I.B. $30 each.

Ford 8N Figurine: This tractor was made by Georgia Marble of Rome, Georgia, in 1997. It was made in the 1/32 scale, and could be purchased for nineteen dollars. Value $20, N.I.B. n/a.

Red Oak™ 8N "Dear Grandpa™": Another item produced by Ertl and Lowell Davis was the Foxfire Farm™ set named, "Dear Grandpa." It is probably classified more as a figurine than a toy, so I originally wasn't going to mention it in this book. However, now that Ford tractors are a thing of the past, this is definitely a beautiful remembrance of the tractor division of the Ford Motor Company. This item was made in China and sold at New Holland dealerships in 1997 for around fifty dollars. Value $25, N.I.B. $50.

Ford Tricycle 8N Figurine: This tractor was made by Popular Imports of China in 1998 in the 1/16 scale. This tractor could be purchased for sixteen dollars. Value $15, N.I.B. n/a.

Pewter 8N Ford: Spec Cast in Dyersville, Iowa, made this 8N Ford in the 1/50 scale. It could be purchased in 1995 for twenty-seven dollars. Value $15, N.I.B. $30.

Precision 8N Ford and Dearborn Plow: The Ford tractor division of the Ford Motor Company built over 1/2 million 8N tractors between 1947 and 1952. A large part of their success could have been because of the attractive appearance of the new red and gray paint scheme, compared to the old solid gray paint found on the 2N and 9N. Farmers had to distinguish between the new 8N and the old Ford Ferguson so the new tractor got the name "Red Belly." Ertl produced the 1/16 scale "Red Belly" in late 1996 in their highly detailed Precision Series for collectors who thought that a hundred dollar bill was reasonable for a well detailed replica.

Mounted behind this 8N is the famous Dearborn two bottom plow. It was also available in late 1996 as an accessory, which could be neatly mounted to the precision 8N. The plow and a Dearborn blade came boxed together and could be purchased in the seventy-five dollar range. Value tractor $100, N.I.B. $200; Value plow $25, N.I.B. n/a.

Precision 8N and Dearborn Blade: Probably the most attractive toy in my Ford collection is the 8N Precision Series by Ertl. This one is identical to the one in the previous photograph except for the Dearborn rear mounted blade. Value tractor and blade $125, N.I.B. $225.

Danbury Mint 8N Ford: This 8N Ford was made in the 1/16 scale by Danbury Mint of Norwalk, Connecticut. It could be purchased in 2000 for $127. Value $100, N.I.B. $127.

8N Ford and Dearborn Wagon: This tractor and wagon was made by Scale Models in 1997 in the 1/8 scale. The tractor and wagon could be purchased for $225. Value Set $100, N.I.B. $225.

Chapter Three
Ford Jubilee Series 1953-1954

1953 Golden Jubilee™: In 1953 Ford Motor Company celebrated its fiftieth anniversary. The Ford and Ferguson lawsuit was settled and the mandatory design changes were underway for the new Golden Jubilee model. The term "Golden Jubilee" refers to the fiftieth anniversary of the Ford Motor Company and jubilee is a biblical term found in Leviticus 25: 11 which refers to a fifty year period. The 1953 Jubilee tractor had a nose medallion, which read, "Golden Jubilee Model 1909-1953." This same medallion could be seen on the 1/12 scale Jubilee toy made of plastic by Product Miniature. It too was made in 1953 of highly detailed plastic and included a three-point hitch that would allow one of these Dearborn two bottom plows made by Slik Toy Company to be attached. Value Tractor $250, N.I.B. $400; Value Plow $250, N.I.B. $350.

1954 Ford NAA (Jubilee): The correct name for the 1954 Ford model tractor is NAA, but today it is still referred to as a Jubilee. The NAA toy was identical to the 1953 Golden Jubilee except for the nose medallion. The 1954 model NAA had stars around the medallion instead of the words "Golden Jubilee Model, 1903-1953." Product Miniature also showed this change in this 1/12 scale plastic model that was also made in 1954. Value $250, N.I.B. $400.

Ertl Golden Jubilee: This 1/16 scale Golden Jubilee was first made by Ertl in 1986. It came in a regular edition that could be purchased for around fifteen dollars and in a collector's edition as seen here for thirty dollars. Value $20, N.I.B. $35.

Jubilee and Combine: The 1/16 scale 1953 Golden Jubilee was made by Ertl in late 1987 and could be purchased for sixteen dollars. It has the plastic wheel rims with front plastic and rubber rear tires. The attached Carter Tru-Scale™ toy is modified to resemble a combine made for Ford by the Woods Brothers. Value Set $75, N.I.B. n/a.

Golden Jubilee "Grandpa's Girl™": In 1997 Lowell Davis and Ertl of Dyersville, Iowa, produced this Fox Fire farm figurine and tractor set called "Grandpa's Girl." Grandpa's name was "Frost" and the little girl's name was "Becky." My New Holland dealer charged thirty-five dollars for this set. Value $20, N.I.B. $35.

Ertl 1953 Golden Jubilee: If you would have told me this photograph was of a full scale Ford Jubilee tractor I would probably have believed it. Seriously speaking, I'm trying to point out the quality and detail on this Precision Series Ford Jubilee manufactured by Ertl of Dyersville, Iowa. This accurately detailed 1/16 scale model was available from the Ford farm implement dealer in time for Christmas in 1997. That is of course, if you could find an extra one hundred dollar bill lying around and you already had your 1997 county taxes paid. Value $100, N.I.B $125.

Franklin Mint 1953 Ford Golden Jubilee: In 1998 Franklin Mint made the 12 scale model of Ford's Fiftieth Anniversary Golden Jubilee. This Ford tractor featured operating steering wheel, opening hood, and operating clutch and brake pedal. There was even an opening toolbox built into the running board as seen on the original tractor. Detail on this model was so accurate there even appeared to have been an electrical wire going to the taillight. This Franklin Mint model could be purchased the easy way of five monthly payment of twenty-seven dollars each. *Background in photograph courtesy of Sue Latham.* Value $75, N.I.B. $145.

Chapter Four
Ford 600 and 700 Series

Ford Comic Book: This is a comic strip type magazine that was given to Ford tractor prospects in the mid 1950s, when the model 600 Ford was new. It demonstrated how a farm family could pay for the Ford Tractor just by the extra income that it could provide. Value: $15.

Ford 600 Tractor and Dearborn Wagon: Ten short years after World War II ended, a young Montana farmer decided to ease his farming burdens with a new Ford 600 tractor and Dearborn Wagon. On the exciting day he purchased this machinery, he was presented with a miniature 1/12 scale replica of the full-scale equipment. The miniature tractor he received that day was this Ford 600 made in 1955 by Product Miniature Company of Milwaukee, Wisconsin. It was constructed of well detailed red and gray plastic and had rubber tires and a steering wheel. It is shown here with its original box. The toy Dearborn flair box wagon he received was made in 1953 by a company known as The Lansing Manufacturing Company, of Lansing, Iowa. This red metal wagon with real rubber tires was made by Lansing under the name of Slik Toy, a name many collectors associate with a nice variety of farm toys.

Well, for some reason or another, these toys escaped the rough hands of child's play. Then, in later years, when the children were grown, none of them seemed to care about dad's toys. So with several acts of swift bidding, and the changing of hands a few times, they have ended up in my collection – with a big kid who believes they are still wonderful toys. Value Wagon $150, N.I.B. $300; Value Tractor $325, N.I.B. $650.

641 Ford Workmaster™: This 641 Workmaster started its life in 1955 as a model 600 plastic toy made by Product Miniature in the 1/12 scale. Somewhere down the road of its hard life in a Midwest sandbox, its hood was lost. The remainder of that 600 ended up in my collection. After several years of searching for a 600 plastic hood and not finding one, I decided to borrow a metal hood from a Scale Models 900. With a couple of 641 Ford decals, this toy once again came to life representing a slightly newer model. Value $75, N.I.B. n/a.

"Toy Tractor Times" Ford 641: *Toy Tractor Times*, a farm toy magazine, from Osage, Iowa, had Ertl make this 641 Workmaster toy to commemorate their 1998 anniversary. The model was made in the 1/16 scale and represented Ford's 4-wheel utility tractor with a 134 cubic inch Red Tiger™ engine. The number four indicated this tractor had a four-speed transmission with PTO and a three-point hitch. The toy had die cast metal rims, special black canopy, and a laser etched logo which read, "The Toy Tractor Times Anniversary 1998." The model sold for forty-eight dollars plus two dollars and fifty cents for shipping and handling. *Photograph background courtesy of Sue Latham.* Value $35, N.I.B. $75.

Ford 621 Workmaster: In November 1999, Ford collectors could buy this nice little 1/16 Ertl scale model of a 621 Workmaster for around twenty-three dollars. The full sized 621 model numbers would represent a four wheel utility tractor with a four speed transmission without a PTO or three point hitche. *Background in photograph courtesy of Sue Latham.* Value $20, N.I.B $25.

Ford 681 Workmaster: This small 1/64 scale tractor from Ertl was first introduced in late 1999 in the red hooded Workmaster trim design. The life sized 681 represented a Ford four wheel utility tractor with Select-O-Speed transmission, non live PTO, and a three point hitch. New Holland dealers offered this toy to their customers for three dollars. Value $3, N.I.B. $5.

771 Select O Speed: This toy Ford tractor started as a Scale Model Ford 900 that was modified by painting the hood red, adding a shift lever, and a set of Select-O-Speed decals which made it resemble a 771 model. This was the tractor Ford used to introduce Select-O-Speed to the American farmer. Value $30, N.I.B. n/a.

Fox Fire Friends™ 771: The 771 Select-O-Speed Ford tractor was made by Ertl in 1998 and manufactured in the 1/16 scale. This toy tractor had the black exhaust and plastic wheel rims, and a cuddly bear that sat in the driver's seat. This toy was purchased for twenty-five dollars. The tractor was made in the Special Edition without the collector's insert. Value $25, N.I.B. $45.

Chapter Five
Ford 900 Series

Product Miniature 900 and Ford Plow: When collecting toys, several things add to their value. If the toy itself was a low production item, if the toy is desirable by collectors, and if the toy is in exceptionally nice condition with its original shipping carton, the seller can make a nice profit.

This 1/12 scale Ford 900 was manufactured around 1955 of high quality plastic by Product Miniature and had everything it takes to drive its value sky high. In junk condition, a farm toy dealer can easily bring home a hundred bucks. My advice to Ford collectors would be, when buying a plastic Ford, pay a little more, and buy a nice one. If you ever decide to sell it, you might not get your money out of a "junker." This does not, however, apply to the Ford 900; if you can find one in any condition, buy it. I have seen very few 900s mint and in the original box. These can usually be obtained by inheriting one, winning the lottery, or by knowing the banker on a first name basis.

Since the first two rules didn't apply to me, I decided to trot down to the First National to see if I could borrow the full amount needed to purchase this toy tractor. When it was my turn to see the loan officer, I sheepishly told him I'd like to buy a 900 Ford Tractor and I asked him if he thought $900 was a good price. He thumbed through his farm tractor price guide, and promptly approved the loan. All was fine for a couple of days, until he drove out to the farm. He needed a photograph of the 900 as it was being used for collateral. I said, "No problem" as I headed down to the toy room and he headed out to the machine shed. I doubt if I will be able to get any more farm loans at that bank. Value $500, N.I.B. $975.

Slik Toy Company manufactured the two-bottom plow around 1955. It was identical to the earlier 1/12 scale Dearborn two-bottom plow except, of course, for the decal. Today you can expect to pay around one hundred dollars for a plow and three hundred if you can find it in its original box. Value $100, N.I.B. $300.

Scale Models Ford 900: In 1991, Scale Models made a nice 900 Ford tractor using the old 4000 Hubley dies. The end product was a new grill that gave the tractor the appearance of the earlier 900. Value $25, N.I.B. $45.

Irwin 901 Ford: Irwin Toy Corporation, originally from New York City, made this plain looking toy tractor in the late 1950s. Close examination of the cast metal 5.5 inch toy distinctly shows a hood, which resembles the styling used on Ford tractors from 1953 through 1961. This toy appears to be a high clearance tractor with the vertical grill and red hood. That design would identify this toy as a 771 Select-O-Speed tractor. However, most toy collectors refer to this toy as the Irwin 901 Ford. This toy can also be found in a plastic variation. Both variations have the Irwin name cast into both sides of the hood. Value $25, N.I.B. $75.

Ford 960: This Ford 960 tractor was made in the mid 1950s, and could be purchased new in the Ozark Mountains for four dollars and ninety-eight cents. This 1/10 scale toy tractor didn't have the "three point hitch" commonly found on the Ford toys of the 1950s and 1960s. It did however have a lift lever, which was used to raise or lower a cultivator (shown attached) or a cultipacker, which were accessory items made for this toy. The "H" found in the nose medallion stood for "Hubley" its manufacturer. a toy company from Lancaster, Pennsylvania. Value Tractor and Implement $125, N.I.B. n/a.

961 Powermaster™ and Three-Bottom Plow: The most common 961 Ford toy is this 1/12 scale Hubley Powermaster in the row crop design. As a second grader in 1958, I was given one of these tractors and a three-bottom plow for my birthday. My dad took me uptown to Carl's Dime Store and he purchased the tractor for two dollars and sixty-nine cents. The plow was an extra dollar. Value Tractor $200, N.I.B. $350; Value Plow $125, N.I.B. $250.

961 Powermaster Wide Front: This 961 Hubley Powermaster is identical to the Powermaster in the previous photograph except for the wide front axle design. Collectors will find this model hard to find especially in its original box. Value $225, N.I.B. $450.

Scale Models 901 Powermaster: The 901 Powermaster was made in the 1/12-scale and looks like the Hubley toy Fords of the late 1950s. Although it was cast from the old Hubley dies, Scale Models made this tractor in 1986. It is referred to as the Show Tractor for the National Farm Toy Show that was held November 8, 1986 in Dyersville, Iowa. This reproduction of the old Hubley Model didn't have a three-point hitch, shift lever, or a power steering arm. This 901 could be purchased for twenty-seven dollars. Value $30. N.I.B. $50.

Gold 961: The gold-plated 1/12 scale tractor made by Hubley, is considered by collectors a hard to find toy and information on this toy tractor is even harder to find. Most collectors refer to this model as a dealer award, which was given to a very limited number of dealers. It was awarded possibly for progress in sales of the actual tractor. At any rate, this model is a miniature copy of the actual gold colored demonstrator tractor found at the local Ford tractor dealerships in 1958. Value $350, N.I.B. $500.

Select-O-Speed and Posthole Digger: This Select-O-Speed toy by Hubley has the row crop or narrow front end design. Attached to the tractor is the Dearborn posthole digger that could be purchased at Carl's Dime Store in 1959 for a dollar bill. Value Tractor $200, N.I.B. $325; Value Digger $200, N.I.B. $350.

Select-O-Speed with Grader Blade: An interesting variation of the Hubley 961 Powermaster is this Select-O-Speed model shown here with its Dearborn grader blade. In the late 1950s Select-O-Speed had a power shift transmission, which was a little ahead of its time as farmers in the late fifties were skeptical of the clutchless transmission. The three-point grader blade made by Slik Toy is hard toy to find, especially in its original box. In 1958, I took a dollar bill to Carl's Dime Store with the intention of buying one, but all I could find was the post hole digger. The post hole digger didn't appeal to me, so I left with the dollar bill still in my hand. Value Tractor $175, N.I.B. $350; Value Blade $200, N.I.B. $375.

901 Powermaster: This 1/64 scale Ford is the 901 Powermaster manufactured by Ertl in 1994. Toy stores sold them for three-dollars. Value $3, N.I.B. $5.

Ertl 901 with Carter Tru-Scale Picker and Wagon: This unusual creature is an Ertl 901 Select-O-Speed Ford Tractor made in 1987 in the 1/16 scale. The rear fenders have been removed to accommodate the TruScale picker, which was modified to resemble a corn picker, built in the 1950s for Ford by the Wood Brothers. The wagon is also a Tru-Scale and should probably have a Dearborn decal to be correct for this trio of Ford implements. Value for set $175, N.I.B. n/a.

Ford 981 and 250 Hay Baler: The special edition 981 Select-O-Speed Ford tractor was found at Ford dealers in 1987 for twenty-five dollars. It was a nice model with cast metal wheel rims, black muffler, vertical exhaust, and full rubber tires. The collector insert on the left side claimed the toy was a "1987 Special Edition Ford 981," but the tractor sported a 901 decal. The baler was once a Carter Tru-Scale , but now resembles the Ford model 250. By the way, the young lady demonstrating the baler for me is a 1994 McDonald's™ Barbie™. Value $100, N.I.B. n/a.

Toy Farmer 901: This 901 Powermaster Ford tractor was made by Ertl in 1986 and manufactured in the 1/16 scale in Dyersville, Iowa. This toy tractor had the gray exhaust and metal cast wheel rims and could be purchased for thirty-eight dollars. The tractor was made in a special edition with the insert that read, "National Farm Toy Show 11-8-86." Value $50, N.I.B. $75.

Regular Edition 901: A regular edition 901 Select-O-Speed Ford tractor was made by Ertl in 1987 and manufactured in the 1/16 scale. This toy tractor had the gray exhaust and plastic wheel rims and could be purchased for fifteen dollars. Value $15. N.I.B. $25.

Collector's Edition 901: The 901 Select-O-Speed Ford tractor was made by Ertl and manufactured in the 1/16 scale in Dyersville, Iowa. This toy tractor had the black exhaust and metal cast wheel rims The tractor was made in the special edition with the collector's insert that read, "1987 Special Edition 981." Its original price was twenty-five dollars. Value $25, N.I.B. $35.

Gold 901 Demonstrator: This 901 Select-O-Speed Ford tractor was made by Ertl in 1994 and manufactured in the 1/16 scale This toy tractor had the black exhaust and metal cast wheel rims and could commonly be purchased for thirty-three dollars. The tractor was made in the special edition with a collector's insert that read, "N.H. Parts Expo Special Edition 1994." Value $35, N.I.B. $45.

Fox Fire Farm's 901: In 1995, Lowell Davis, a painter and sculpture from Carthage, Missouri, teamed up with the Ertl Company to produce this 1/16 scale 901 Ford toy with a figurine sitting in the seat. The driver's name was Jim Babcock. This set sold at New Holland dealerships for forty dollars. Value $35, N.I.B. $50.

Firestone™ 901 Ford: This 901 Select-O-Speed Ford tractor was made by Ertl in 1998 and manufactured in the 1/16 scale. This toy tractor had the black exhaust and cast metal wheel rims and could be obtained by purchasing a set of tractor tires. Value $35, N.I.B. $75.

Chapter Six
Ford 1000 and 2000 Series

Ford 1710 with Roll Over Protection System: This Ford 1710 was made in the 1/16 scale by Ertl of Dyersville, Iowa. It could be purchased in 1986 for thirteen dollars. Value $25, N.I.B. $35.

Ford 1710 Special Edition: This Ford 1710 was made in the 1/16 scale by Ertl and it could be purchased in 1985 for thirty dollars. It had a collector's insert which read, "Ford 1710, 1985." Value $30, N.I.B. $40.

Ford 1920 Compact Tractor: This Ford 1920 was made in the 1/16 scale by Scale Models and could be purchased in 1988 for thirty dollars. Value $20, N.I.B. $30.

Scale Models 2000: In 1986 I purchased this Ford model 2000 for thirty dollars. It had the row crop front end, 2000 Ford decal and the traditional solid blue hood. Scale Models made this toy in the 1/12 scale from the old Hubley dies. Cast into the left fender were the words, "1986 Toy Show Tractor." Value $30, N.I.B. $75.

Britains Ford 2120: Britains, a toy company in England made this nice 1/32 scale Ford 2120. This toy represents Ford's smaller line of tractors. It is shown here with a front wheel assist and a cab. A cab is not commonly found in the United States on the smaller sized tractors, however in England new tractors are required to have a roll over protection system or a cab. In this case a cab was chosen. This tractor was purchased in 1992 for sixteen dollars. Value $20, N.I.B. $30.

Danbury Mint Ford 2000: This 2000 Ford has a long list of precision crafted items which include a hood that opened, PTO shaft, valve stems, starter, dash gauges, reversing rear wheels, and the list goes on and on. I purchased this 1/16 scale 2000 by mail for $127. Value $100, N.I.B. $135.

Chapter Seven
Ford 4000 and 5000 Series

4000 Dime Store Tractor: Collectors refer to this unusual Hubley tractor as a dime store 4000. This red and gray 4000 was an inexpensive model not found at Ford tractor dealerships. The toy could be easily found in dime stores, or five and ten cent stores, as they were commonly referred to in the 1960s. The tractor was less expensive because it didn't have solid rubber rear tires, 4000 hood decals, Select-O-Speed dash shifter, or the power steering arm. These items were all found on the other 1/12 scale Hubley 4000 Ford tractors. *Background of photograph courtesy of Sue Latham.* Value $300, N.I.B. $500.

Red Wide Front 4000 Ford: This unusual Hubley 4000 Ford tractor is red and gray instead of the traditional blue and gray. Why? Were the early 4000s using leftover Select-O-Speed bodies? Did someone at Hubley make a mistake? Did Hubley have a lot of red paint to get rid of? Some collectors say this variation was also a dimestore edition, made in a more expensive version. Whatever the reason, it is a very interesting and unusual Ford toy. *Photograph courtesy of Jim Newman* Value $250, N.I.B. $450.

4000 Narrow Front: One of the hardest to find Hubley 4000s today is this narrow front or row crop version. It has blue and gray trim, steer-able front wheels, solid rubber tires, three point hitch, 4000 hood decals, and the Select-O-Speed shifter on the dash. Value $175, N.I.B. $350.

4000 Wide Front: In 1962 Ford tractor changed to the blue and gray paint scheme for their farm tractors. This was the first Hubley tractor I remember seeing that reflected this change. I was in the sixth grade in 1962 and I would often see this toy at Gibson's discount center. A sixth grader was just too old to ask for toy tractors, besides the 961 I received in 1958 was still in good shape. This blue and gray Hubley 4000 is the most common and easiest variation to find today. It had the 4000 decal, three point hitch, solid rubber tires and the Select-O-Speed shifter on the dash. At the discount store it could be purchased for less than four dollars. Value $150, N.I.B. $300.

4000 LPG Ford: This photograph shows a modified version of a 4000 Ford, equipped to burn LP gas (liquid petroleum). This fuel was cheaper to use than gasoline, but it didn't produce the horsepower that gasoline could provide. This type of fuel wasn't very popular with most farmers and finally stepped aside to make way for the diesel engine. Value $150, N.I.B. n/a.

Scale Models Row Crop 4000: Scale Models bought the old dies from Hubley and produced this 4000 Ford in the 1/12 scale in 1982. It didn't have a three point hitch, shift lever, or the power steering arm. It was a nice toy at the price of twenty-five dollars. Value $25, N.I.B. $45.

Scale Models Wide Front 4000 : This later variation manufactured in 1995 of the Scale Models 4000 had a wide front end and the plastic rear wheel hubs. It sold for thirty-five dollars at New Holland dealerships. Value $20, N.I.B. $45.

1962 Ford 4000: In my opinion, the little 1/64 scale blue and gray 4000 Ford Utility is one of the nicest looking tractors Ford ever produced. The only model of this tractor in my collection is this example that originally cost three dollars and was made in China by Ertl. Value $3, N.I.B. $5

Early Split Grill 4000: In 1965 Ford Motor Company introduced to farmers its totally new "World Tractor." This was a combination of the American Ford tractor, and the English Fordson tractor. The American farmers got a new three cylinder engine, the English farmer got a shorter Ford name and farm kids got a nice 1/12 scale Ford toy with a suggested five dollar retail price. This was Ertl's first Ford tractor and was only produced from 1965 to 1968. The photo shows a very early split grill 4000 with rare cast metal rear wheel rims. This split grill tractor is also seen in its original thirty five year old shrink wrap box. Value $150, N.I.B. $300.

Split Grill 4000 with Plastic Wheel Rims: This 1/12 scale tractor was manufactured by Ertl in 1968 and could be purchased for five dollars. It had the plastic rear wheel rims not found on the earlier Ford 4000. *Background in photograph courtesy of Sue Latham.* Value $100, N.I.B. $200.

Rare 4000 Ford: This is probably one of rarest Ford toys in existence. It has an unusual red Ford 4000 decal on a blue background which is similar to the one found on the early 8000 by Ertl. I've only seen a few of these tractors in my years of collecting. The toy is a very early 1/12 scale Ertl Ford flat grill manufactured around 1968. The value of this toy is hard to determine due to its extreme rarity. Value $350, N.I.B. $700.

Ford 4000 With One Piece Grill: Ford engineers designed a new one piece grill to be used on the 1968 products. Ertl followed suit in the 1/12 scale to make a replica of this tractor to reflect that change. Two other changes appeared in this model that showed concern for child safety. The first was the soft muffler that replaced the plastic one found on the split grill model. Another safety feature was the new flat top three point hitch lever introduced some time after production of this model had already begun. I bought one of these 4000s in 1969 at the Missouri State Fair for five dollars and seventy-five cents. Value $65, N.I.B. $125.

Ford 4600 with White Wheel Centers: New decals, fenders, front and rear wheels were found on this 1/12 scale Ford toy. A slightly modified body casting that allowed the muffler to be moved forward away from the fuel tank filler spout was used on the full sized 4600 as well as on this Ertl Replica.

This model was made with a heavier front axle and the front wheels had the white inside wheel centers not found on later 4600s. Rumor has it that some of the early 4600s were produced with the old round fenders and large 4000 type wheels. However, remember these can easily be made with a neat decal change so check the date on the bottom and don't pay big bucks for a repaint. This toy Ford could be found with a gray or black round muffler or the more common gray or black oval muffler. These tractors could be purchased with or without the three point hitch for about eighteen dollars. Value $35, N.I.B. $55.

Ford 4600 with Black Wheel Centers: For those of you who collect the variations, this 4600 has the more common dark inside front wheel centers. The tractor was made with or without the three point hitch. It was the last version in the Ertl 4000 Ford series, which lasted almost thirty-five years. Value $35, N.I.B. $55.

Ford 4610: When Ford tractor division discontinued the 4600 for the improved 4610 model, I hoped Ford would have Ertl make this change in the 1/12 or 1/16 scale. The model that appeared however was a new 1/16 scale 7710 with front wheel assist. To satisfy my disappointment, I took a hack saw in one hand and with some sales literature in the other, I figured if Ford could make a 4610 tractor in full scale, I might be able to make one in the smaller scale. The round nose likeness was carved from a 2 x 4 piece of wood. The decals were made from Ertl 7710 decals and the air cleaner came from a Hubley 6000. The end result was rewarding but not as good as Ertl would have done at Mr. Ford's request. Value $35, N.I.B n/a.

County Super™ 4-754: In 1984, Dave Sharp of Indiana made a nice conversion of the County Super 4-754 Model from an Ertl Ford 4000. County of England modified the real 4000 Ford tractor for use in the wet climate. The unique thing about the County version was a drive shaft coming from each rear axel housing to the front wheels. The cost for Dave's County conversion was around one hundred dollars. Value $175, N.I.B. n/a.

Britains 100 Anniversary: In 1993, W.F. Britains Company of England celebrated their 100 anniversary. A 1/32 scale Ford 5610 toy tractor in a very limited edition came painted silver with gold trim and rainbow decals. To commemorate this event, a Britains decal with a 100 was used instead of a Ford 5610 decal. This model in the United States was around thirty dollars, but the money was probably easier to find than the tractor. Value $50, N.I.B. $80.

Ford 5000: This little 1/64 scale Ford Super Major 5000 was made in 1994 by Ertl. This toy sold in the three dollar price range. Value $3, N.I.P $5.

Ford 5000 with Side Scoop: Corgi toys of Wales made this nice split grill 5000 Ford tractor with side mounted scoop. It was manufactured in the 1/43 scale of die-cast metal from 1969 to 1972 and was usually available at major airports or fine department stores here in the United States. Value $45, N.I.B. $100.

Ford 5000 Super Major: Corgi of England made this Ford 5000 in the 1/43 scale. It could be purchased for five dollars. Value $25, N.I.B. $25.

Ford 5610 with Black Decal: This Ford 5610 was made in the 1/32 scale by Britains in England. It could be purchased in 1987 for ten dollars. *Background in photograph courtesy of Mary McCleery.* Value $15, N.I.B. $25.

Ertl 5000 Ford: The little 5000 Ford tractor shown in this photograph was manufactured by Ertl. The full-scale tractor that this toy represents was made from 1968 to 1975. This 1/64 scale tractor was purchased for three dollars. Value $3, N.I.P $5.

Ford Super Major 5000: Britains made this nice little 1/32 scale Ford Super Major 5000 English version of the split grill tractor around 1965. In the United States, this tractor was referred to as a Ford 5000. In England it was known as the Ford 5000 Super Major. Notice the English style fenders, which would prevent the operator from accidentally getting a foot or leg caught in the rear wheel. Value $50, N.I.B. $150.

Ford Precision 5000: In the past years Ertl would usually introduce a Ford tractor in the precision series in December. These tractors would make a nice Christmas gift for the Ford collector, however, this one wasn't available until late January 2000, but it was well worth the wait. It was Ford number seven in Ertl's Precision Series and looked huge next to the precision 8N and Jubilee. The tractor's blue paint and great detail made a nice looking 5000 model tractor, which helped me forget about the one hundred and ten dollars I paid for it. Value $75, N.I.B. $125.

Majorette 5000 Ford: Majorette made this little 1/55 scale Ford 5000 around 1972 in France. They were painted red or metallic blue and were available with a variety of implements. Value $15, N.I.B. $25.

Galanite Ford 5000: The little 1/43 scale Ford tractor seen in this photograph was made around 1970 by Galanite of Sweden. This toy was often referred to as a 4000, but it distinctly shows a 4-cylindar diesel engine which in real life would represent the 5000 model.

The toy could be found in a variety of bright colors and was usually found with a four wheel wagon in matching colors. Value $15, N.I.B. n/a.

Ford 5000 Plastic Model Kit: Britains made this Ford 5000 model kit in the 1/43 scale. It could be purchased in the late 1960s for one dollar. Value $25, N.I.B. $125.

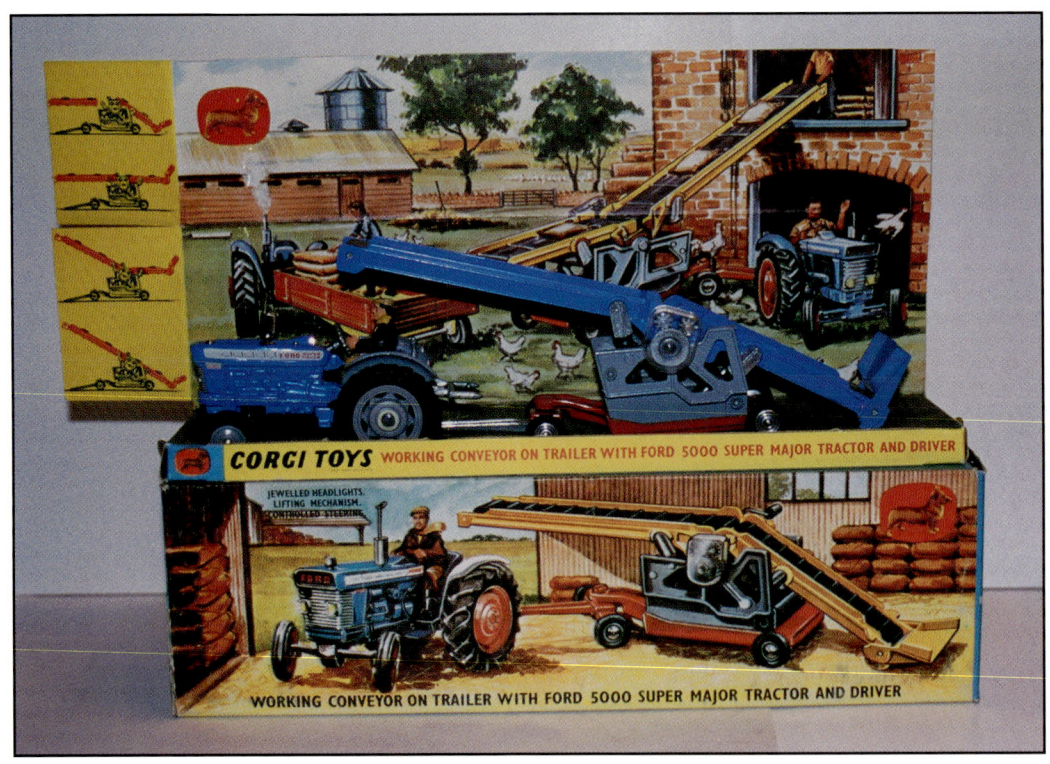

Ford 5000 with Conveyor: Corgi Toys made this nice conveyor set mounted on a trailer in the mid 1960s. The toy included a 1/43 scale Ford 5000 Super Major tractor with man, in an English farm scene box. *Courtesy of John Hill.* Value $75, N.I.B. $150.

Chapter Eight
Ford 6000 and 7000 Series

Red 6000 Diesel: In 1961, the Ford Motor Company introduced its first production tractor with a six-cylinder engine, the model 6000. The 1/12 scale toy manufactured by Hubley made its first appearance in 1963. This toy had the aluminum muffler, rubber tires, and a super detailed three-point hitch that could be attached to the earlier Hubley machinery of the Ford 961 and 4000 models. The tractors side decal read,"6000 Diesel" in the matching red color scheme. *Background in photograph courtesy of Sue Latham.* Value $250, N.I.B. $475.

Blue 6000 Diesel: A year after the red 6000 toy diesel was introduced, Hubley made this identical blue 6000 Diesel to follow suit of the color change in the full scale tractors. The blue 6000 Diesel also had the aluminum muffler, rubber rear tires, and a detailed three-point hitch. The side decal was blue and read, "Ford 6000 Diesel" however some earlier versions were manufactured with the red 6000 Diesel Decal. Value $200, N.I.B. $425.

Ford Commander 6000: In 1965 Ford tractor division introduced a more powerful 6000 tractor. Most of the improvements on this tractor couldn't be seen, but the changes that could be seen included the new split grill design. This gave the tractor a new appearance and a different name, the Commander 6000. The 6000 now had hollow, plastic rear tires that didn't add to the appearance of the toy. I purchased this Commander in 1969 for nine dollars. The tractor in the photograph has an aluminum replacement muffler and an air cleaner. It is pictured with the Hubley three bottom plow. Value Tractor $125, N.I.B. $215; Value Plow $100, N.I.B. $175.

Gabriel Commander 6000: This photograph shows the fourth Ford 6000. The Hubley division, now owned by the Gabriel Toy Company of Lancaster, Pennsylvania made this toy. This 6000 has the split grill, three point hitch, and hollow plastic tires. Gone forever are the muffler and the air cleaner. This toy Ford could be purchased in the late 1970s for ten dollars. Value $75, N.I.B. $150.

Last Gabriel 6000: Gabriel also made this variation of the Ford 6000. It is probably the most common 6000 to be found today. It has the split grill design, hollow plastic rear tires and no muffler or air cleaner. Gabriel also made this tractor without a three point hitch. The only machinery that could be attached to this 6000 was a blue cultivator or the blue cultipacker which was similar to the one made by Hubley for the 1/10 scale Ford 960. Value $65, N.I.B $150.

Scale Models 6000: Scale Models made this Ford 6000 in the 1/12 scale. It could be purchased in 1998 for forty-five dollars. Value $35, N.I.B. $55.

6600 by Olti: An unusual toy here in the United States is this Ford 6600 made in the 1/12 scale by Strike (Olti) of South Africa. This toy is made of stamped steel with a three point hitch and somewhat crude engine design from glued on plastic engine parts. The tractor was made also in a yellow industrial version with a front-end loader. Both versions are hard to find in the United States. I purchased this tractor in 1991 for fifty dollars. Value $100, N.I.B. n/a.

Britains 6600 Ford Tractors: Both Britains tractors in this photograph are 6600 Fords made in the 1/32 scale. The tractor on the left has the cab, muffler, air cleaner, Ford decals, and a driver. It could be purchased in 1981 for ten dollars. A less expensive model shown on the right could be found in 1982. Even the Ford name was absent from this brown and orange tractor. I paid eight dollars for this hard to find toy. Value $25 each, N.I.B. $45 each.

Ford 6600 with Roll Over Protection System (R.O.P.S.): This Ford 6600 was made in the 1/32 scale by Britains. It could be purchased in 1985 for nine dollars. Value $20, N.I.B. $45.

Blue Ford 6610: This blue 1/64 scale tractor manufactured by Maisto had a chrome engine compartment and the Ford decal was found on both hood sides. It could be purchased in 1996 for one dollar. Value $1.00, N.I.B. $2.00.

Red Ford 6610: The 1/64 scale red tractor by Maisto has a chrome engine compartment and the Ford decal is found on both hood sides. It could be purchased in 1996 for one dollar. Value $1.00, N.I.B. $2.00.

Blue Ford 6610 with Loader: Maisto made this 1/64 scale tractor which I purchased in 1996 for one dollar. The blue tractor has a chrome engine compartment and the Ford decal is found on both hood sides. This tractor has R.O.P.S. instead of the cab. Value $1.00, N.I.B $2.00.

Yellow Tonka™ with Loader: The Maisto Company also made this 1/64 scale tractor which I purchased for one dollar in 1999. This tractor has a chrome engine compartment as found on the earlier 6610s with the exception of the Tonka decal instead of the Ford decal. It also has the R.O.P.S. and the front-end loader. Value $1.00, N.I.B. $2.00.

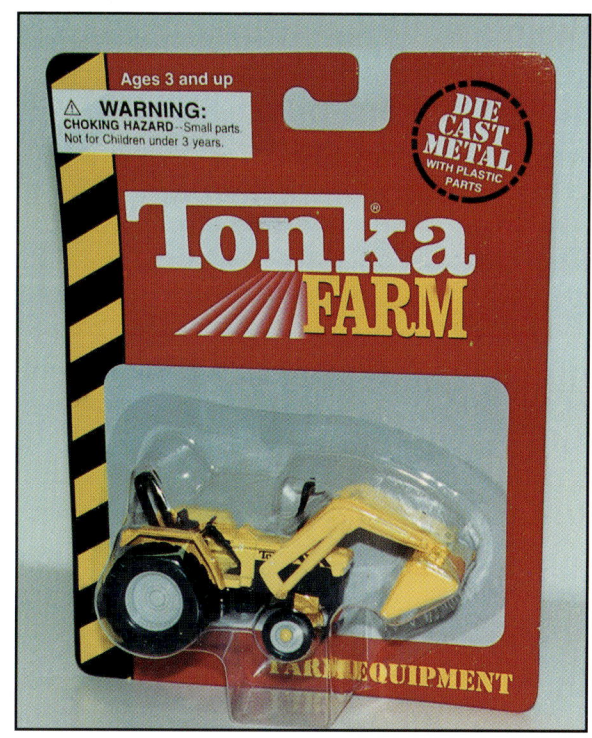

Ford 6635: This Ford 6635 was made in the 1/32 scale by Britains. It could be purchased in 1997 for sixteen dollars. Value $16, N.I.B. $20.

Ford 7700: Some of the Ertl 1/12 scale Ford toy tractors were produced from the same body casting with the six cylinder engine. The 7700 should have had a four cylinder engine, however, we must remember toy tractors were children's toys and detail didn't outweigh economics. This photograph shows the 7700, which was first made in 1977. This tractor could be purchased for fifteen dollars. Value $125, N.I.B. $225.

Ford 7710 Classic: This Ford 7710 Classic was made in the 1/16 scale by Ertl and it had a collector's insert which read "Ford Classic 7710." It was purchased in 1982 for forty dollars. Value $50, N.I.B. $65.

Ford 7710: This toy 7710 made by Ertl was first available as a farm toy in the summer of 1982. It was Ertl's second 1/16 scale Ford toy and I welcomed the new smaller size. With all the 1/12 scale Ford toys I had accumulated over the years, display space was beginning to be a problem. These early 7710s did not have a rivet in the suitcase weights as seen on the later 7710s. The toy sold new for twelve dollars. Value $25, N.I.B. $35.

Toy Farmer 7710 Ford: This 7710 toy Ford tractor was manufactured by Ertl to commemorate the 6th National Farm Toy Show. The show was held November 11-12, 1983. Special features of the 7710 Ford included the larger rear tires, larger front assist tires, gloss gray fenders R.O.P.S., and the addition of the 8000 style air cleaner. Another special feature of this toy was the special engraving found on the left side of the operator's platform which read "National Farm Toy Show Dyersville, Iowa, 11-12-1983" and a "Zeke" medallion attached to the front grill. Value $200, N.I.B. $350.

7710 Wheatland™: Although Ford never made a Wheatland tractor in the 7710 series, I thought this would be a nice looking addition to my collection. I started with a set of rear fenders from a Minneapolis Moline™ G1000, added dual Hubley 4000 headlights to the fenders, installed a swinging drawbar, heavy duty air cleaner, aluminum muffler, Firestone diamond tread tires, extra large rear tires, and ended up with a tractor that looks almost as good as a 5020 John Deere™. Value $90, N.I.B. n/a.

7710 with Cab: Ertl did not put a cab on their 1/16 scale 7710, but a friend of mine modified an Ertl 7710 by installing a Scale Models TW25 cab. The toy looked great especially when connected to a Bush Hog™ rotary cutter. *Photograph courtesy of John Hill.* Value Tractor $125, N.I.B. n/a.

7710 Series II: This variation of the 1/16 scale 7710 by Ertl was made in 1986. It was now labeled by Ford as the 7710 Series II. This toy version was basically the same toy as the earlier 7710s except for the new black and red identification decal and the addition of the Series II decal. This toy sold new at most Ford New Holland dealerships for thirteen dollars. Value $25, N.I.B. $35.

Ford 7610: Lonestar of Great Britain made the Ford 7610 in the 1/32 scale. It could be purchased in 1984 for ten dollars. Value $15, N.I.B. $25.

Ford 7610 with Front and Rear Duals: Lonestar also made this Ford 7610 in the 32 scale. It could be purchased in 1986 for nine dollars. *Background in photograph courtesy of Mary McCleery.* Value $15, N.I.B. $25.

Ford 7710 with Power Assist: Britains made this 1/32 scale 7710 Ford toy. It could be purchased in 1984 for eight dollars. Value $20, N.I.B. $30.

Chapter Nine
Ford 8000 and 9000 Series

Early 8000: Over the past year of trying to photograph and list variations of toy tractor models, someone will inform me of a new one that I didn't know about. A large part of these variations are in the gray or blue center rear wheels or the three point or non three point toys. I won't try to mention all of these because wheel and three point hitches can be switched, so the variations mentioned in this book are of non restored toys that appear to have come from the factory in their original form.

The first 8000 in the 1/12 scale from the Ertl company is this blue background 8000, which is one of the most difficult to find in non restored condition. It was made for such a short period of time that it didn't even appear in Ertl's toy catalogue. According to information in the Ertl magazine, Replica, all of the particular variations of the 8000 had the three point hitch. This Ford toy was available in 1969 for eight dollars. Value $125, N.I.B. $250.

Ford 8000 on White Background: This second 8000 variation has the "Ford 8000" on a white decal. It first appeared in the 1969 Ertl catalogue in the 1/12 scale. According to the Ertl magazine, Replica, it was only manufactured with the blue center rear wheels and a three point hitch. This toy was available in 1969 for eleven dollars. Value $75, N.I.B. $150.

8000 Narrow Front: This photograph shows an early 1/12 scale tractor with modified tricycle front end. The Ford tractor division made the actual tractor for a very short period of time and it was designed to handle the two row corn pickers, which were becoming less popular in the 1960s. The tricycle type toy wasn't produced by any toy manufacturer instead it had to be custom built by a collector. *Photograph courtesy of Steve Drake.* Value $100, N.I.B, n/a.

Ford 8000 Pulling Tractor: Tractor pulls were fast becoming popular in the late 1960s. The Ford tractor in the photograph is a modified version of an 8000. In the early days of tractor pulls, the tractor was unhooked from a plow, hay rake, or mower and driven from the field to the school's agricultural class tractor pull. These pulls were usually fundraisers. The only preparation was an engine tune up and maybe a fresh tank of gasoline, certainly not the version seen in the photograph. The modifications to this 8000 would make this a very appealing addition to a collection. *Photograph courtesy of John Hill.* Value $125, N.I.B. n/a.

Large 8000 Variation: Ertl's third major variation came out in 1970 and had a nice looking decal. This decal also appeared on the full sized Ford tractor as well as on the 1/12 scale toy. This 8000 was made with or without out the three point hitch. This toy could be purchased for eleven dollars. Value $70, N.I.B. $100.

Plastic 8600 Ford: Processed Plastics made this 8600 in a variety of color and decal combinations. It was made totally of plastic and could be purchased in the mid 1980s at discount stores for approximately two dollars. Value $15, N.I.B. n/a.

Ford 8600: This Ford toy first appeared for sale in 1973, with decals much like those of the 8000 toy tractor. It could be found in several combinations featuring a tractor with or without the three point hitch, blue metal center rear wheels, or smaller gray center plastic rear wheels. This 8600 toy could also be found in a farm set, which included the big blue wagon. The tractor was manufactured by Ertl and usually found in the eighteen dollar range. Value $50, N.I.B. $100.

Ford 8700 with Canopy: Polistil of Italy made this Ford 8700 in the 1/43 scale. It could be purchased 1983 for ten dollars. Value $20, N.I.B. $30.

Ford 8630 Power Shift: This was an extremely nice 1/32 scale toy for a ten dollar bill. This tractor was made in 1991 and came with a front-end loader and several attachments. The tractor had a working three point type hitch and front wheel assist. The tractor and attachments were packaged in a neat Styrofoam™ lined box. If you happen to find one I encourage you to make a quick purchase, as this set was made for only a short period of time. Value $25, N.I.B. $45.

1/32 scale Ford 8630: This Ertl 8630 Ford toy tractor was made around 1993 and could be purchased for approximately ten dollars. It came with the regular hitch, power assist front wheels and had the black decal. Value $10, N.I.B. $15.

Ford 8630 with Loader: New Ray of China made this Ford 8630 in the 1/32 scale in 1999. It could be purchased for twenty dollars. Value $20, N.I.B. $30.

Ford 8830 and 8730: In 1990 the TW Series tractors were discontinued and the four digit numbering system was again used to identify the Ford tractor models. The tractor on the left is an 8830 with front wheel assist, which had a price tag in early 1991 of thirty-one dollars. The 8730 tractor on the right was made in 1990 in the row crop design by Scale Models and could be purchased for twenty six dollars. A big brother of the 8730 with dual wheels was also available for an extra three dollars. Value $30 each. N.I.B. $60 each.

8830 with Windows: This is a modified 1/16 scale of the 8830 with windows made by Scale Models The original wheels were replaced with those from an Ertl 8340 Power Star. The Scale Models Haybine™ was made in 1990 in the 1/16 scale and sold for sixteen dollars. Value Tractor $35, N.I.B. n/a. Value Haybine $25, N.I.B. $35.

Britains Ford 8560: This Britains toy in the 1/32 scale was manufactured without an engine. No, I don't know why, except to say it was an early production model as the later 8560s were made with crude engine detail. This toy was purchased in 1996 for sixteen dollars. Value $20, N.I.B. $30.

Britains 8560 with Flotation Tires: This nicely detailed Ford New Holland 8560 with engine detail and attractive front and rear flotation tires was made in 1998. It was a nice 1/32 scale tractor for twenty-one dollars. Value $25, N.I.B. $30.

1/32 scale Ford 8670: This Ertl 8670 Ford toy tractor was made around 1994 and could be purchased for approximately ten dollars. It came with the regular hitch, power assist front wheels, and it had the black decal. Hitched to the tractor is an Ertl minimum till plow. It was made in 1983 and could be purchased for three dollars. Value Tractor $15, N.I.B. $25; Value Plow $10, N.I.B. $25.

Ford 8770 with 660 Round Baler: This Ford 8770 appears to be identical to the 8970. It would be except for the model number variation, and the addition of the New Holland emblem in the place of the old Ford emblem on the nose of the tractor. This 1/16 scale toy made by Spec Cast of Dyersville, Iowa, has an insert cast into the tractor's left front frame which read "8770, 2nd Edition". The tractor was made in the regular steer edition and was available to anyone who had an extra seventy-one dollars.

The 660 Ford New Holland round baler pictured was made by Scale Models in 1991. It came as a collector's edition that could be purchased for twenty dollars. It had a collector's insert on the left side which read, "Parts Mart 1991" Value Tractor $75, N.I.B. $100; Baler $25, N.I.B. $50.

Collector Series 8870: There weren't many Ford collector series tractors made in the 1/64 scale. This tractor represents the Ford Genesis™ 8870, and was made by Ertl. A unique feature of these toys was the Plexiglas windows and the collector's insert found on the cab top which read, "Genesis Collector 1995." The model shown is attached to a Ford New Holland model 660 round baler, which was made by Scale Models in 1992. Both toys could be were priced in the seven-dollar range. Value Tractor $10, N.I.B. $15; Value Baler $10, N.I.B. $25.

End of an Era Ford: This special edition Ford 8870 tractor was only available to collectors for maybe ten or fifteen minutes. Not many showed up at the farm toy shows, so if you purchased one at the seventy-five dollar price, hang on to it. Spec Cast made this toy in 1996 with a nicely detailed engine that was accessible from its tilt up hood. Value $90, N.I.B. $135.

Ford 8970 with Oval Emblem: Spec Cast made the 1/16 scale Ford collector's tractor in 1994. It featured Super Steer front wheel drive. This was a front axle that moved sideways to allow the front wheels to make a tighter turn. The model had a collectors insert on the tractor's left side which read, "8970 First Edition 1994." This tractor would probably be considered a model that won't be found in many sandboxes because of the seventy-five dollar price tag. For this price, the buyer had a tractor with a hood that lifted to allow access to a nicely detailed turbo charged six cylinder engine. Value $75, N.I.B. $90.

Pewter 8970 Ford: This tractor was made in the 1/50 scale by Spec Cast in 1995. It could be purchased for thirty-six dollars. Value $30, N.I.B. $38.

Ford 8970: Jouef of Italy made this Ford tractor in the 1/25 scale. It could be purchased in 1998 for twenty-six dollars. *Background in photograph courtesy of Mary McCleery.* Value $25, N.I.B. $35.

Ertl produced this 1/8 scale Ford 8000 pedal tractor as a fundraiser for the National Farm Toy Museum in Dyersville, Iowa. Included in the forty nine dollar price you received an early 8000 pedal tractor with white steering wheel and a pass for two of your friends or family members to visit the National Farm Toy Museum. This toy had a collector's insert which read "NFTM, SERIES 3, #3, 1999." Value $25, N.I.B. $50.

Ford 9000: This 9000 wasn't produced as a toy but started its life as an Ertl 9600. When the kids were finished playing with it, I purchased it, made a 9000 decal, and painted the cab gloss gray so it would resemble the 1969 model 9000. Today a more realistic 9000 decals are available from farm toy parts dealers. Value $110, N.I.B. n/a.

Ford 9600: This 9600 in the 1/12 scale was made in 1974 by Ertl and came with or without the three-point hitch. The dual rear wheels made this toy appear to be wider than it was long. This toy commonly had the gray center plastic rear wheels; however, I have seen this toy with what appeared to be factory mounted dual blue centered metal rear wheels. Value $100, N.I.B. $200; with blue center rear wheels and three point hitch Value $150, N.I.B. $250.

1/64 scale Dome Wheel 9700: If you like the 9700 Ford toy in the 1/12 scale but don't have an extra quarter acre to display them, this little 1/64 scale tractor might just do the trick. It was made in Hong Kong by the Ertl Company around 1978. This one has three variations not found on later models. It has the silver grill, a cast in power take-off housing, and the hard to find Ford style dome rear wheels. I remember giving these little toys as birthday gifts in the late seventies. They were nice, inexpensive gifts, but now they bring as much as three hundred dollars. Value $50, N.I.P. $250.

1/64 scale 9700 with PTO Casting: This 9700 is much easier to find and easier on the credit card to purchase than the 9700 with dome wheels. It was made in 1980, but it doesn't have the PTO housing or the dome wheels. The grill is now black as on the full size 9700. In 1980, this Ertl toy could be found in stores for less than one dollar. Value $15, N.I.P. $45.

Ford 9700: In 1977, Ertl made this 9700 Ford toy tractor in the 1/12 scale. It was the same toy as the 7700 except for the decals and the addition of the dual rear wheels. This toy was made from 1977 until 1982 and was priced slightly less than twenty dollars. Value $75, N.I.B. $125.

Chapter Ten
Ford TW Series

Ford TW5 (Blue Decal): This blue decal Ertl TW5 was made from 1985 until 1986, when Ford tractors changed to a black background decal. This blue decal TW5 had a production change not found in the previous models. Its front wheels were now held on by a push type cap identical to the one used to hold the rear wheels in place. This new Ford also had a new higher price tag of twenty-seven dollars. Value $65, N.I.B. $85.

1/32 scale Ford TW5: This Ertl TW5 Ford toy tractor was made around 1988 and could be purchased new in the seven dollar range. It came with the chrome three-point hitch, power assist front wheels, and had the black decal. Value $15, N.I.B. $25.

1/32 scale Ford TW5: The Ertl TW5 Ford toy tractor was made around 1990, and could be purchased new in the eight-dollar range. It came with the black three-point hitch, power assist front wheels, and it also had the black decal. Value $10, N.I.B. $15.

Ford TW5 (Black Decal): If you didn't run out of money buying the big 1/12 scale Ford toys, you would surely run out of room to display them. At twenty-five dollars or more for these big tractors, I was beginning to wish Ertl would start making Ford Farm toys in the smaller 1/16 scale. The new TW5 with its new black decal and flat roof made one nice looking Ford toy. Ertl made this toy from 1986 until 1990. Value $65, N.I.B. $100.

Ford TW10: In 1981, this 1/12 scale Ertl TW 10 replaced the 7700 as the tractor with the single rear wheels. This tractor could be bought from 1981 to 1983 for around twenty dollars. Value $65, N.I.B. $85.

1/32 scale Ford TW 15: This Ertl TW15 Ford toy tractor was made around 1986 and could be purchased new in the eight-dollar range. It came with the regular hitch, power assist front wheels, and the black decal. Value $10, N.I.B. $15.

Ford TW15 (Blue Decal): As usual, a big brother model was made for the junior farmer that needed a little more horse power and wanted the dual rear wheels. The answer had to be the blue decal Ertl TW15, made in 1986. Identical to the TW5, both were made with the new push on front wheel caps. In the 1/12 scale, this toy sold new for around twenty eight dollars. Value $65, N.I.B. $85.

Ford TW15 (Black Decal):This is the one that I was waiting for, the last Ertl 1/12 scale Ford tractor. The Ford toys were made in the 1/12 scale by request of Henry Ford. The 8-N was basically a small tractor, designed for railroad boxcar shipping. Mr. Ford didn't want children to grow up remembering his tractors as being small, so he insisted the toys be made in the larger 1/12 scale. The black decal TW15 with its new flat top roof was made from 1986 until 1990. This large toy had a large price tag of thirty dollars. That was double the price of the 7700 in 1977, but I'm glad to own one because now dealers will ask over a hundred dollars for one in mint condition. Value $65, N.I.B. $100.

1/64 scale TW20: Ford tractor division changed the 9700 to the TW20 in 1979. This photo shows a 1/64 scale TW20 made by Ertl. It had the square rear axle housing like the early 9700 toy. Value $5, N.I.P. $10.

Ford TW20: This Ford TW20 was made in the 1/32 scale by Britains It could be purchased in 1981 for twelve dollars. *Background in photograph courtesy of Mary McCleery* Value $20, N.I.B. $35.

Ford TW20: This 1/12 scale TW20 is big brother to the TW10. It was made in the dual wheel version only and carried a price tag of twenty-five dollars. Ertl made this toy from 1981 until 1983. Value $75, N.I.B. $100.

TW20 Pedal Tractor and Trailer: I don't collect the Ford pedal tractors, so when Ertl began producing this 1/8 scale model of a TW20 pedal tractor and trailer, I briefly glanced at them in a magazine article. Six months later, I saw this set at a Ford New Holland dealership and I was impressed with the quality of the set. I picked one up and carried it to the counter, where I found myself thirty dollars poorer. Value $30, N.I.B. $35.

Ford TW25: Ertl began making this blue decal Ford toy tractor with single rear wheels in mid 1983. This toy was not a true replica of the actual tractor due to the discrepancy in the length of the hood. The actual tractor had a longer hood but this was not reflected in the toy tractor. My guess is that the Ford Tractor Division requested a decal change. This toy isn't rare in the collecting business, but it is somewhat hard to find. If you bought this tractor when it first came out, you probably paid around twenty dollars for it. Value $65, N.I.B. $85.

Ford New Holland Mixer and TW25: This grinder mixer was made in 1987 by Ertl and could be bought for fifteen dollars. It is connected to a scale model Ford TW25, which was made special for the Ford New Holland dealer meeting, which was held in Tennessee in 1989. I purchased the tractor from my Ford dealer for twenty-five dollars that same year. Value Tractor $25, N.I.B. $35; Value Mixer $25, N.I.B. $35.

TW Series II™: The TW25 Series II was an improved model of the old TW25. This toy shows the Series II decal below the Ford name and came equipped with dual wheels. This toy was manufactured by Scale Models. The cab top has an inscription which reads "Coming on Strong, Boston 1990," which designates this toy tractor as a 1990 Ford New Holland dealer meet edition. This toy tractor could be owned for twenty-eight dollars. Value $25, N.I.B. $35.

Ford POW-R-PULL™ Tractors: This little 1/64 scale Ford by Ertl came to the United States in two totally different variations that were made in two different countries. The Ford on the left was made around 1985 in Korea. The Ford toy on the right, was from China, and manufactured around 1987. They both came in the POW-R-PULL series, which had a spring-loaded motor. The two toy TW35s were totally different including decals, tires, wheels, exhaust stack, and air cleaner. It is probably safe to say that no items on the two toys were identical. Both variations were easy to find in the late 1980's and could be purchased for about three dollars each. Value $5 each, N.I.P. $10 each.

Ford TW35 with F.W.A.: Siku in West Germany made this TW35 Ford in 1/32 scale. It could be purchased in 1985 for thirteen dollars. This tractor has its identification decal installed backwards; it should read Ford TW35. Value $15, N.I.B. $30.

Ford TW35: This 1/12 scale Ertl TW35 is, of course, big brother to the TW25s. It was made in 1983 and discontinued by early 1985. Like the full size TW25, this toy should have had a much longer hood and was replaced in 1985 by the Ford toy model TW15. This toy may also be somewhat hard to find today in good original condition, partly due to the fact that very few collectors were willing to pay the twenty-five turnips it took to buy this big dual wheel toy. Value $65, N.I.B. $85.

Battery operated TW35: This Ford battery operated TW35 was made in the 1/32 scale by Britains. It could be purchased in 1987 for ten dollars. Value $20, N.I.B. $35.

Above: Buddy L. Ford: This unusual Ford tractor was made in 1982 of plastic and tin. It could be found in an assortment of colors and closely resembled Ford's TW Series of tractors manufactured in the 1980s. Buddy L, a toy company known for its nice toy trucks made this tractor in Hong Kong. This toy could usually be found in discount stores for less than four dollars. Value $10, N.I.B. $20.

Left: 50ML Pacesetter Ford: In 1983 Pacesetter of Bardstown, Kentucky, made these little Ford whiskey bottle tractors in approximately the 1/32 scale. They came in two different variations; one variation was ready to go with a full 50ml tank of vodka and the other was for us Baptist collectors, with an empty tank. Value $35, N.I.B. $70.

750ML Pacesetter Ford: The old saying, "Things aren't always as they appear to be," may be correct in this situation. This appears to be a 1/16 scale toy tractor, but when it was new in 1983 it contained 750ml of vodka. This Ford tractor manufactured by Pacesetter was the third in a series. I mysteriously only paid a dollar for this tractor which contained a half tank of vodka. I sometimes ask myself, "Did the owner not like vodka?" "Did he not like Ford tractors?" "Why was he so anxious to get rid of it?" I will never know. Value $40, N.I.B. $75.

Plasto Ford Pull Toy: This generic toy tractor was made in the mid 1980s by Plasto of Finland. This plastic child's pull toy tractor is complete with Ford decals and done up in the traditional Ford blue and white colors. Value $10, N.I.B. n/a.

Chapter Eleven
Ford Power Star Series

Ford 8340 Collector Edition: The first Power Star™ toy I found was the 8340 in the collector's edition, which was made in 1992. In the 1/16 scale, it had front wheel assist, spin out style rear wheels, three step plates, straight pipe exhaust, cab with dash side panel doors, and engine side panel covers. A collector's insert was found on the left side which read, "Collector Edition 1992" on the first line and "8340 4WD" on the second line. The Ford toy cost thirty-seven dollars. Value $40, N.I.B. $50.

7740 Collector's Edition: this Power Star 7740 Collector's Edition was purchased in 1992. It had the two-wheel drive design with the spin out style rear wheels and three-step plates with straight pipe type exhaust. This toy has the cab with dash side panel doors and the collectors insert on the left side, which read, "Collector's Edition 1992, 7740 2 WD tractor." This model sold for forty-two dollars. Value $45, N.I.B. $65.

Ford 8240 with Cab: The third Power Star model was the 1/16 scale, two-wheel drive made by Ertl in late 1993. It had the old 4600 Ford style dome rear wheels with narrow tires, two step plates, straight pipe type exhaust, cab with dash side access, and engine side panels. This toy was not a collector's edition, so the cost was a little more reasonable at twenty-seven dollars. Value $30, N.I.B. $50.

Ford 6640 with Cab and Front Wheel Assist: The fourth Power Star also came out in late 1993. It was the 6640 model with front wheel assist. It also had the old 4600 Ford dome style wheels, however this one had wider tires. It came with step plates, straight pipe exhaust, a cab with dash side access door, and no engine side covers. This one could be purchased for twenty-seven dollars. Value $30, N.I.B. $50.

Ford Power Star 5640 Collector's Edition: This 1/16 scale Power Star was the 5640, which I purchased from my Ford dealer in June 1994 for forty-five dollars, It was made in the collector's series and had an insert which read, "Collector's Edition 1994, 5640 2WD Tractor." This was a nice toy with Ertl's big spin out rear wheels, three step plates, R.O.P.S., and a new muffler type exhaust. This exhaust stack was used on the actual 5640 tractor because the non-turbo charged engine was nosier and required a muffler to lower the noise it produced. The main casting of this toy was the same as the 7740 without a cab. This toy doesn't have the dash side access doors, but it does have the casting for the longer decals. Value $50, N.I.B. $65.

Ford 7740: The fifth Power Star Ford toy had a different body casting. It was not made with the dash side access doors, and did not have an indentation in the hood casting for a longer decal, which to my knowledge was never made. This 7740 came with front wheel assist, three step plates, 4600 style dome rear wheels, and wide tires. It also had the common straight pipe exhaust and cost thirty-six dollars. Value $35, N.I.B. $50.

Ford 7840: The Power Star 7840 came in the collector's edition with the usual left hand collector's insert which read, "Collector's Edition 1994, 7840 4W.D. Tractor." The tractor came in the front wheel assist design with the spin out rear wheels, three step plates, straight pipe type exhaust, dash side access doors, and the engine side panels. It sold new for forty-five dollars. Value $50, N.I.B. $65.

Ford 8340: The 8340 Power Star came out in early 1994. It was the two-wheel drive model in the non-collector's edition. As seen here, it came with the roll over protection system, and has the narrow 4600 style rear wheels, two step plates and dash access side doors. I purchased this toy for thirty-six dollars. Value $35, N.I.B. $50.

Ford 7840: The Ford 7840 was made in the row crop front wheel design with single rear wheels. This tractor came with a cab and could be purchased on the Ford New Holland Card for three dollars. Value $3, N.I.P. $5.

Ford 8340 with Dual Rear Wheels: The Ford 8340 was made in the row crop front wheel design with dual rear wheels. This 1/64 scale tractor came with a cab and could be purchased on the Ford New Holland Card for three dollars. Value $3. N.I.P. $5.

Ford 5640 with Loader: The Ford 5640 with loader was made in the row crop, front wheel design with single rear wheels. This tractor came with roll over protection system and could be purchased on the Ford New Holland Card in 1992 for four dollars. Value $3, N.I.P. $5.

Ford 7740 with Loader: The Ford 7740 with loader was made in the row crop design with single rear wheels. This tractor came with a cab and could be purchased on the Ford New Holland Card for four dollars in 1993. Value $4, N.I.P. $7.

Ford 6640: The Ford 6640 was made in the row crop, front wheel design with single rear wheels. This tractor came with roll over protection and could be purchased on the New Holland Card for three dollars in 1993. Value $3. N.I.P. $5.

Ford 8240 4WD: The Ford 8240 4WD was made with power assist, front wheel design and single rear wheels. This 1/64 scale tractor came with a cab and could be purchased on the Ford New Holland Card for three dollars. Value $6, N.I.P. $10.

Ford 7740 4WD: The Ford 7740 4WD was made with power assist front wheel design and single rear wheels. This 1/64 scale tractor came with a roll over protection system and could be purchased on the Ford New Holland Card for three dollars. Value $6. N.I.P. $10.

Chapter Twelve
Ford Industrial Series

Industrial Ford 1841: In the 1950s, Cragstan, a toy company located in Japan, made this industrial Ford 1841 loader backhoe in approximately the 1/12 scale. It was an extra nice battery operated toy made of stamped tin with a lighted engine compartment that showed moving engine components. This toy came in the traditional Ford red and gray paint scheme as found on the farm series tractors. Value $125, N.I.B. $250.

Industrial Ford 4000: A later version of the battery operated industrial toy Ford, was this model 4000 HD also made in the 1/12 scale. Its manufacturer, Cragstan, stayed with the traditional color scheme of the full sized tractors. This red and yellow backhoe-loader tractor also had the lighted engine compartment and battery operated front-end loader. A crank found in the operator's control area operated the well detailed backhoe. Value $125, N.I.B. $250.

Red and Yellow Ford 4040 Industrial: This Cragstan loader backhoe is almost identical to the others I have mentioned except for the red and yellow color scheme. This photograph shows this toy's exceptionally nice shipping carton. This model was also made during the 1950s. *Photograph courtesy of John Hill* Value $125, N.I.B. $250.

Blue and Yellow Ford 4000 HD Industrial: This blue and yellow Cragstan 4000HD industrial was almost identical to the earlier versions except for the new blue and yellow paint scheme. This paint combination was used on this model from 1962 to 1964. The remote control unit had a steering wheel to guide the toy from left to right, a forward button, a reverse button, and two more buttons that raised or lowered the front-end loader. Like the earlier models, the backhoe was crank operated. Value $125, N.I.B. $250.

Ford Remote Control Forklift: This early Cragstan Forklift was made in the 1960s. The toy could travel in forward or reverse, left and right, and the forklift could be raised and lowered. All of these functions could be operated by a hand held remote unit that operated on two D size flashlight batteries. *Photograph courtesy of John Hill.* Value $200, N.I.B. $450.

Ford Number 21 Forklift: Another Cragstan battery operated forklift was also manufactured in the early 1960s. Functions of this toy were forward and reverse, which is operated from the right side shifter. This lever also served as the on/off switch. The drive and forklift maneuvers is controlled by the left side lever, which also activated the blinking light. The driver is made of a dyed, flexible rubber. The Cragstan forklift could be found in the Montgomery Wards Christmas catalog for under ten dollars. Value $250. N.I.B. $500.

Ford 4400 Industrial: This 4400 was made around 1972 using the same body casting as found on the farm toy version of the 4000. The industrial style decals and industrial yellow paint scheme seem to be the only differences. The 4400 in the photograph shows the earliest version with the round three point hitch control lever, later, a flat top style lever was used with child safety in mind. Both versions have been out of production for over twenty-five years and they are seldom seen at toy shows today. When buying one of these Ford 4400s, be certain it has the original yellow plastic wheel rims and not the painted ones. Value $125, N.I.B. $250.

Industrial 8400: After the kids get married and move off the farm, the question arises, "What should be done with those big Ertl 8000s the kids have out grown?" If you haven't given them to the church garage sale you can turn one of them into a beautiful blue and yellow Industrial 8400. This toy was never produced by Ertl, but the decal set could be purchased from various toy parts suppliers. *Photograph courtesy of John Hill.* Value $175, N.I.B. n/a.

4000 Industrial Power Unit: An interesting model was the 4000 industrial power unit that was made using a hood and engine from a Scale Models 4000 tractor. Value $75, N.I.B. n/a.

Ford 7710 Britains Industrial: This 1/32 scale toy by Britains was made in an industrial version for a very short time. It was sold in the Autoway Series™ in 1984. The blade and auger were made to fit any Britains tractor. The Ford tractor cost ten dollars while the blade and auger were in the five-dollar range. Value Tractor $25, N.I.B. $75; Value Blade and Auger $5 each, N.I.B. $10 each.

Britains Industrial 1884: In England, County converted the Ford tractor to front wheel assist to better suit their climate. Britains made this County 1884 in an industrial yellow toy in 1991. It was a well-detailed, good-looking toy in the 1/32 scale for the eighteen dollars it cost here in the United States. Value $25, N.I.B. $75,

Britains Motorized 1884: After the Industrial 1884 County was available here in the United States, a cousin appeared at toy shows for a very short time. So short, in fact, that this example is the only one I have ever seen. It came in the traditional 1/32 scale like the industrial yellow version mentioned earlier, except this one was white in color and had a battery powered motor. This toy required a small AA size battery installed neatly under the removable hood. With the addition of the electric motor, this toy cost an additional seven dollars over the price of the yellow industrial. I do not know why this toy came out in the white version Value $25, N.I.B. $75.

Early 7500 Ford Loader with Backhoe: Ertl made these 1/12 scale loader-backhoes from 1972 to the early 1990s. Because of their size and price, they were passed over by the collectors, but the kids loved them. I bought the 7500 for fifteen dollars in 1975. It came with a black rear bucket and teeth. The early 7500s also had the small front loader hydraulic cylinders. Value $25, N.I.B. $65.

7500 Front Loader with Backhoe: Ertl made these 1/12 scale loader-backhoes from 1972 into the early 1990s. Again due to their large size and limited space of collectors, they were passed over. But now they are sought after by collectors and the new-in-the-box versions are hard to find. I purchased the 7500 for fifteen dollars in 1975. It came with a black rear bucket and teeth. Value $25, N.I.B. $65.

750 Ford Loader with backhoe: Ertl made these 1/12 scale loader-backhoes from 1974 into the early 1990s. The 750 could be purchased for twenty-five dollars in 1981. It came with a yellow rear bucket and teeth. Value $25, N.I.B. $65.

755 Ford Loader with Backhoe: Ertl made these 1/12 scale loader-backhoes from 1974 in the early 1990s. This toy was purchased for thirty-six dollars in 1984. It came with a yellow rear bucket without teeth. Value $25, N.I.B. $65.

755A Ford Loader with Backhoe: This is another 1/12 scale loader-backhoe manufactured by Ertl from 1974 into the early 1990s. I purchased the 755A for 43 dollars in 1989. It came with a yellow rear bucket without teeth. Value $35, N.I.B. $65.

Ford 555 Twenty Piece Construction Set: This loader-backhoe was made by Ertl in 1985 in the 1/32 scale and could be purchased for eighteen dollars. Value $20, N.I.B. $45.

Ford Backhoe: Ertl made this 1/64 scale backhoe in 1990s. It could be purchased for five dollars. Value $3, N.I.B. $10.

Road Construction Set: I purchased this road construction gift set in 1997 for twenty five dollars. It was the same 1/64 scale loader backhoe as shown in previous photographs. The only difference was the set pictured included a video. Included in this set is a John Deere model 544 wheel loader. Oh well, I guess every Ford collection needs at least one John Deere tractor. Value $10, N.I.B. $35.

Pewter Backhoe: One of the last Ford backhoes I purchased was in May of 1997. This is a 1/55 scale backhoe by Ertl. It had the New Holland name, but the rear trencher boom arm still labeled this tractor as a "Ford Model 555E." At fifty-seven dollars it was somewhat expensive, but toy collectors knew the Ford logo would soon disappear completely from the farm and industrial tractor. Value $50, N.I.B. $75.

Ford CL 25 Skid Steer: My parents bought this toy for me as a Christmas gift in 1983. I was thirty-three years old and just as proud of it as when I received my first Ford toy in 1958. This toy was made by Ertl and could be purchased for eleven dollars. Value $25, N.I.B. $40.

New Holland Skid Steer Loader: This toy was made in the 1/16 scale by Ertl. It could be purchased in 1987 for twenty dollars. Value $20, N.I.B. $30.

New Holland Skid Steer Loader Special Edition: This skid steer loader was made in the 1/16 scale by Ertl. It could be purchased in 1986 for twenty-five dollars. It had a collector's insert which read, "Special Edition 1986." Value $25, N.I.B. $50.

Ford New Holland 665: This was the last toy skid loader I added to my collection before the Ford emblem was dropped from the New Holland name. Ertl made this nice little 1/32 scale loader around 1994 and collectors could purchase this toy for less than twelve dollars. Value $12, N.I.B. $20.

Chapter Thirteen
Ford Articulated Four Wheel Drive

1/32 scale FW60s: Ertl made this trio of FW60s in the 1980s. The first toy on the left is a regular edition made in 1981, which could commonly be purchased for around seventeen dollars. The center tractor in the photo was made of plastic and came in a set of four different models. In 1983, the set sold for seventy-five dollars. The last FW60 on the right has a special insert found on the left front frame, which labels it as a "Ford Classic FW60." A serial number is found on the right side and only 1000 of these toys were made. Tractor on the left Value $20, N.I.B. $45; tractor in the center Value $25, N.I.B. $45; tractor on the right Value $75, N.I.B. $125.

Ford Versatile 1/32 scale: Shortly after Ford New Holland acquired Versatile, these toys appeared on Ford dealer's shelves. The red Versatile 1156 Designation 6 on the left, was made in the 1/32 scale. In 1986 it sold in the twenty-five dollar price range. The blue Ford Versatile 1156 on the right is a "Parts Mart" edition made in 1990 and could be purchased for twenty-six dollars. Scale Models of Dyersville, Iowa, made both toys. Value red Ford Versatile $30, N.I.B. $60; Value blue Ford Versatile $25, N.I.B. $50.

Bi-directional Set: This set was made in the 1/32 scale to commemorate the 1989 Ford New Holland dealer meet, which was held that year in Nashville, Tennessee. The tractor on the left is a Ford Versatile bi-directional model 256 and the Ford Versatile on the right is model 846. Both had a dealer meet inscription with date cast in the cab top. The price of set usually ranged between thirty-five to fifty dollars. Value $35, N.I.B. $75.

Gold Nashville Dealer Meet Tractor: Some Ford New Holland dealers who attended their 1989 dealer meeting in Nashville, Tennessee, received a gold plated Ford 846 Versatile tractor. They were made by Scale Models in the 1/64 scale and weren't available to the public. The only way to obtain one of these toys is to find a Ford dealer who was willing to part with his tractor. Engraved on the cab top is "1998 Nashville Dealer Meeting." Value $100, N.I.B. $200.

Some Ford New Holland dealers who attended their 1990 dealer meeting in Boston, Massachusetts, received a gold plated Ford 876 Versatile tractor. They were made by Scale Models in the 1/64 scale and were not available to the public. Engraved on the cab top is "Coming on Strong, Boston 1990." Value $100, N.I.B. $200.

Ford 9880: Scale Models made this 1/16 scale 9880 Ford tractor in 1996. It came with spacer duals all the way around. It was only available in the Ford version for a very short time. The price for this model was over the one hundred dollar mark. Value $85, N.I.B. $125.

Chapter Fourteen
Ford Lawn and Garden

Ford LGT 145 with Trailer: In 1984 I purchased this Ertl 1/12 scale lawn tractor and matching lawn cart. The cost of this toy was fourteen dollars. Value $25, N.I.B. $50.

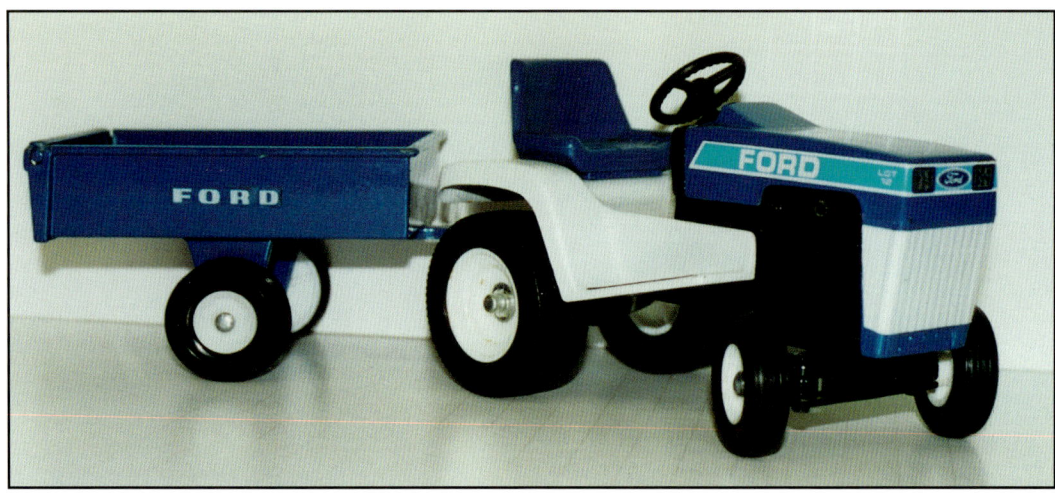

Ford LGT with Trailer: This 1/12 scale Ertl lawn tractor and matching trailer was available in 1984. It could be purchased for twelve dollars. Value $15, N.I.B. $25.

Ford LGT 12 Special Edition: Ertl manufactured a special edition 1/12 scale lawn tractor that was available in 1984 and could be purchased for eighteen dollars. It had a collector's insert on the deck below the steering wheel which read, "Special Edition, March 1984." *Background in photograph courtesy of Mary McCleery.* Value $25, N.I.B. $45.

Ford LGT 18H: This lawn tractor was manufactured in 1994 by Scale Models in the 1/16 scale. It came without a matching lawn cart and could be purchased for fourteen dollars. Value $15, N.I.B. $20.

Ford GT 95: This 1/16 scale lawn tractor without a matching cart was manufactured by Scale Models in 1996. It could be purchased for twenty-one dollars. Value $15, N.I.B. $25.

Chapter Fifteen
Ford Machinery

Dearborn Wagon: Slik Toy of Lansing, Iowa, made this Dearborn wagon in the 1/12 scale. It could be purchased in 1955 for two dollars. Value $75, N.I.B. $300.

Dearborn Two Bottom Plow: The first time I remember seeing one of these 1/12 scale Dearborn two-bottom plows was in 1957. It was toy day at school and all the first graders in my class brought their favorite toy. A friend brought a 1953 Product Miniature Ford Jubilee toy tractor with one of these Dearborn plows attached to the three-point hitch. "I sure would like to have a Ford Tractor and plow," I told my dad that evening after dinner. I got a Ford tractor the next year for Christmas, but the plow was no longer available and could not be found until forty-two years later. Slik Toys, sometimes referred to as Lansing Manufacturing, made this Dearborn two-bottom plow for the Ford toy tractor in 1953 and 1954. Value $175, N.I.B. $350.

Ford Tow-Bottom Plow: Slik toys made this Ford plow in the 1/12 scale. It could be purchased in 1957 for one dollar. This plow was made to be used on the Product Miniature model 600 and 900 Ford Toy tractor. Value $225, N.I.B. $450.

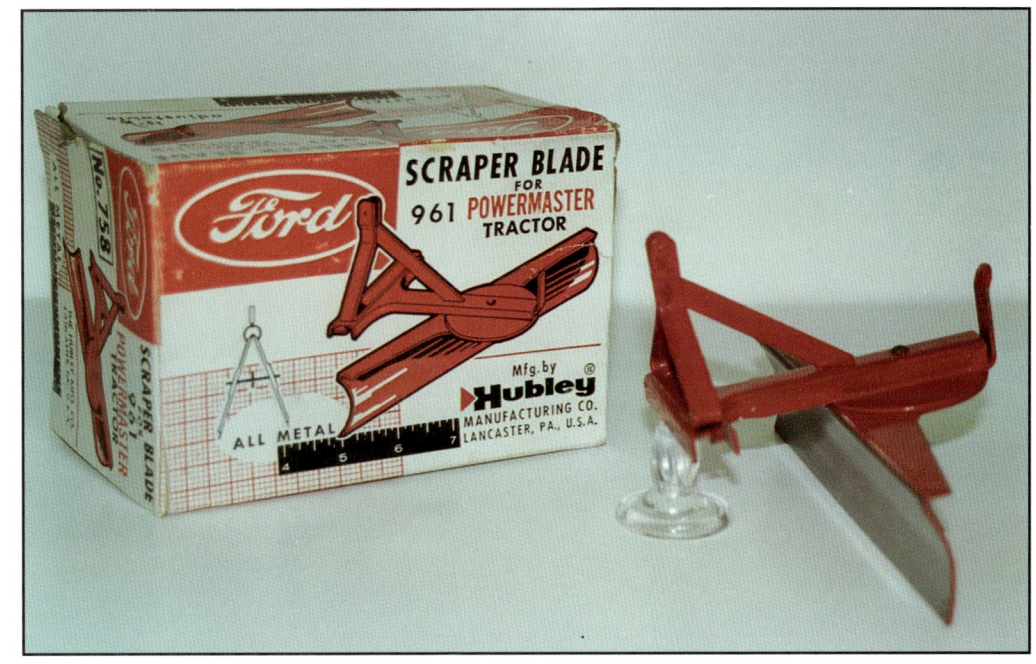

Hubley Scraper Blade: In the summer of 1961, I took a one-dollar bill uptown to Carl's Dime Store with the intention of purchasing one of these Hubley grader blades. You can imagine the disappointment I felt when I discovered it had been sold. The only piece of machinery available to fit my 961 was a posthole digger. Since I didn't have the desire to own a digger, I went back home with my one-dollar bill. I figured I could buy a blade later. I just didn't realize it would take thirty-nine years. This one however, cost me considerably more than one dollar. Value $225, N.I.B. $425.

Hubley Posthole digger: The Hubley posthole digger shown with this Golden Jubilee was made in the late 1950s. It was to be mounted to a 961 Ford toy tractor. Carl's Dime store sold this item in 1960 for a dollar bill, and of course you had to pay two cents sales tax. Value $200, N.I.B. $400.

What is it?: If your guess was an Ertl Precision corn picker painted red to resemble the actual Ford picker found in the late 1950s, then you were correct. A 1/16 scale picker was mounted to a 1/12 scale 961. The project wasn't easy, but with the help of a set of rear wheels from an Ertl 4600 Ford, the set looked wonderful. *Photograph courtesy of John Hill.* Value $500, N.I.B. n/a.

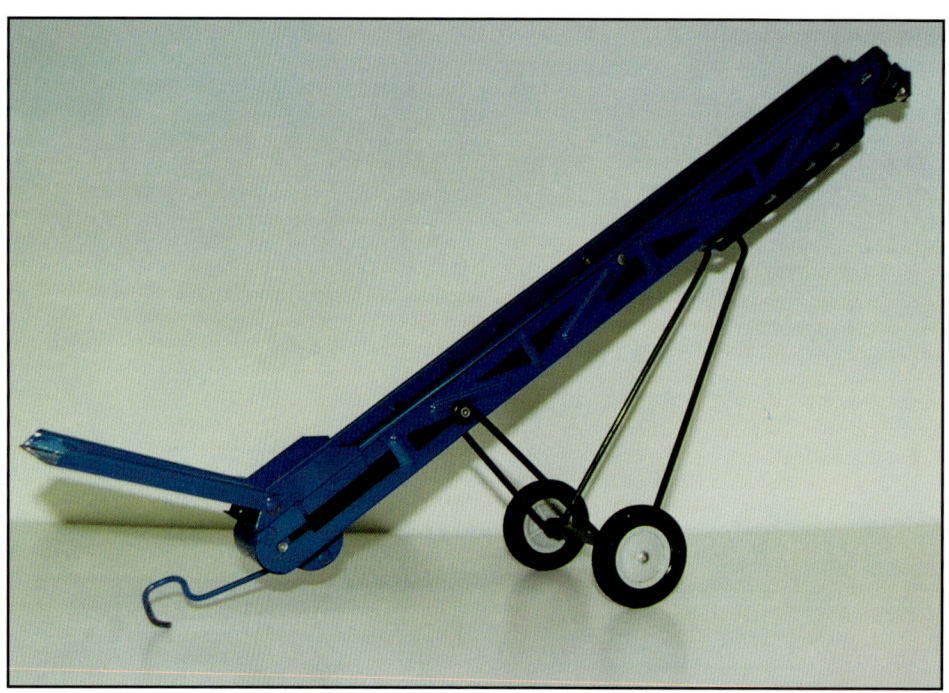

Ford Blue Elevator: I purchased this Ford blue elevator in 1988 at a local farm supply store for seven dollars. It was a generic toy found in a variety of implement colors. Ertl produced this toy at their plant in Dyersville, Iowa. Value $15, N.I.B. $30.

Ford Forage Harvesters: In 1986 Ford Tractor purchased Sperry New Holland, forming the Ford New Holland Company. Within a year, Ford also acquired Versatile Manufacturing Limited of Canada. This photograph shows some of the first toys produced after this merger. The four-wheel drive tractor is a model 836 Ford Versatile that was made by Scale Models in 1988. This 1/64 scale toy could be purchased at most Ford New Holland dealers for six dollars. The forage harvester and wagon are products of the Ertl Company and could be found at a variety of stores in 1987 for two dollars each. The forage wagon has the early gray canopy, which was changed to yellow in 1988. Value for set $20, N.I.P. $40.

Late New Holland Wagon: This photograph shows the later style New Holland 1/64 scale forage wagon with its yellow canopy. It was available in 1988 as a Ford New Holland toy and could be purchased on the yellow card for less than two dollars. Value $5, N.I.P. $12.

Ford New Holland Mower Conditioner: Shortly after Ford purchased New Holland, this little mower conditioner was available for approximately two dollars. They were made in the 1/64 scale and were seen in 1987 in this yellow package bearing the Ertl trademark. Value $3, N.I.P. $10.

New Holland 660 Baler: This 1/64 scale Models baler sold for six dollars in 1993. Value $6, N.I.P. $15.

Gold 50th Anniversary Baler: This baler was manufactured by Ertl in 1991. The selling price was seven dollars. Value $3, N.I.P. $15.

Ford New Holland Baler: This 1/64 scale baler made by the Ertl was available in 1989 for two dollars. Value $3, N.I.P. $10.

Ford Six Bottom Plow: The six bottom plow shown was manufactured by Ertl. It was available in Ford dealerships in 1990 and could be purchased for two dollars. Value $10, N.I.P. $20.

Ford New Holland Hay Rake: This nice little 1/64 scale hay rake was available in 1994 for under three dollars at most stores. This one in the Ford New Holland package was the same rake that was found at John Deere dealerships with green and yellow paint and John Deere packaging. Value $3, N.I.P. $10.

Ford New Holland Box Spreader: If you were buying the 1/64 scale toys in 1993, you probably saw these New Holland Model 145 spreaders made by Ertl. I thought they looked great because the casting was identical to the full sized machine and not just another generic toy painted to resemble the Ford line. Value $3, N.I.P. $10.

!/32 scale Ford TW15: This Ford toy tractor was made around 1985 and could be purchased for seven dollars. It came with the regular hitch, row crop front wheels, and it had the blue decal. Hitched to the tractor is an Ertl winged disc made by the Ertl Company. It was available in 1983 for three dollars. Value Tractor $10, N.I.B. $15; Value Disc $15, N.I.B. $20.

Ford Versatile 2000 Combine: This combine was made in the 1/64 scale by Scale Models. It could be purchased in 1990 for four dollars. Value $5, N.I.P. $10.

Regular Edition TR96: The sandbox model combine was made in the 1/32 scale and could be purchased for twenty-seven dollars. It was available at discount stores or Ford dealerships beginning in 1987. Value $25, N.I.B. $40.

Collectors Edition TR96: This collectors' edition combine is almost identical to the sandbox edition except for the addition of the TR96 chrome decal, the twin spreaders, and the collector insert which labeled this toy as a "New Holland TR96 Special Edition." This Ford New Holland toy commonly sold somewhere in the forty-dollar range. Value $30, $N.I.B $50.

TR 97: The photograph shows the last variation of the Ford New Holland combine. This combine was basically the same casting as the Ertl 1/32 scale TR 96 model. The only two noticeable differences are the new black TR 97 decal and the price increase from twenty seven to thirty four dollars. This combine was available in stores beginning in 1994. Value $25, N.I.B. $40.

Ford NH TR 96 Combine: This New Holland Combine was made in the !/64 scale by the Ertl Company. It came in the old style box and could be purchased in 1990 for nine dollars. Value $10, N.I.B. $20.

Customized Ford NH TR 96 Combine: This custom-built New Holland combine was made in the 1/64 scale and could be purchased in 1990 for nine dollars. It was modified by adding dual wheels to the front and single power assist wheels to the rear axle. Value $20, N.I.B. n/a.

Ford NH TR 96 Combine: A new style box containing this New Holland combine caught the eye of many children in 1993. The combine was manufactured in the 1/64 scale and it was available in 1993 for ten dollars. Value $10, N.I.B. $15.

Ford NH TR 97 Combine: The New Holland combine available in 1994 also came in the new style box. The 1/64 scale combine manufactured by Ertl could be purchased in 1994 for ten dollars. Value $9, N.I.B. $12.

Chapter Sixteen
Ford Farm Sets

Early "N" Series Farm Set: One of the first Ford toy farm sets available in the United States was the Hubley set made in the early 1950s. It consisted of a plastic 1/32 scale Ford "N" series tractor, cultipacker, and a two wheeled trailer. *Photograph courtesy of John Hill.* Value $200, N.I.B. $475.

960 Farm Set: In 1957 Hubley produced a farm set in a super big 1/10 scale. The tractor implements were a wagon with rubber stakes, cultivator, and a cultipacker. *Photograph courtesy of John Hill.* Value $250, N.I.B. $475.

961 Farm Set: Pictured in the photograph is a nice Hubley 961. It was available to the small "carpet farmers" in 1958. The set consisted of a 961 gear shift tractor, a three bottom plow, two wheel trailer, and a crate containing chickens and ducks. *Photograph courtesy of John Hill.* Value $300, N.I.B. $500.

Select-O-Speed Farm Set: In 1962 Hubley introduced a farm set that was identical to earlier sets with the exception of a Select-O-Speed tractor and a yellow farm scene box. This set was identified as farm set number fifty-seven and was identical to the deluxe number sixty-eight farm set, except the deluxe set featured a blade and post auger. *Photograph Courtesy of John Hill.* Value Number 57 Farm Set $250, N.I.B. $500. Number 68 Farm Set $500, N.I.B. $975.

Orange Hover 5000 Farm Set: This farm set was made in the late 1960s by Hover of Hong Kong. The battery-operated tractor came with a red and yellow trailer with a sign on the side that read "James Farm." The farm set came complete with a driver, and four cows. *Photograph courtesy of John Hill.* Value $100, N.I.B. $200.

Blue Hover 5000 Farm Set: Very similar to the farm set shown in the previous photo is this 15 inch long farm tractor, trailer unit. This toy was made in the more realistic blue and white color scheme as seen on the actual Ford tractors of the late 1960s era. *Photograph Courtesy of John Hill.* Value $100, N.I.B. $200.

Flat Grill 4000 Farm Set: This Ertl 4000 set was available in 1972. It consisted of a tractor, wagon, disc, and a four bottom plow. The 4000 tractors were rated in the three to four plow range, but they were seldom seen in Missouri pulling a four bottom plow. *Photograph courtesy of John Hill.* Value $75, N.I.B. $225.

Blue 6000 Diesel Farm Set: The implements shown would be included in a Hubley 6000 Diesel Farm Set. The set included a blue 6000 diesel tractor with removable draw bar, cultivator, cultipacker, and the stake wagon. This wagon had the early "T" type hitch, which connected in a special slot in the tractor's PTO area. Value $300, N.I.B. $650

TW 20 Farm Set: The 1/64 scale farm set by Ertl cost five dollars in the early 1980s. It consisted of a TW20 tractor, round baler, six bottom plow, and a wagon. All the machinery in this set had the tongue hooks. Value $12, N.I.B. $25.

Black Decal TW 35 Farm Set: The farm set shown was manufactured by Ertl in the 1/64 scale. The set consisted of a black decal TW35 Ford tractor, round baler, wing disc, and a wagon. The set could be purchased for seven dollars. Value $15, N.I.B. $25.

Black Decal Six Piece Farm Set: A TW25 Ford tractor with a black decal, front loader, TW35 Ford tractor with a black decal, wing disc, round baler, gravity feed wagon, and a six bottom plow was included in this 1/64 scale farm set. It was manufactured in 1989 by the Ertl Company and cost fourteen dollars. Value $18, N.I.B. $30.

Chapter Seventeen
Ford Dealer Trucks

Matchbox Ford Tractor Trailer: Lesney of Great Britain produced this Ford tractor-trailer unit with a load of three split grill farm tractors in the early 1970s. The matchbox farm tractors were all blue with yellow wheels and are a little harder to find than the yellow and blue industrial version. I bought this toy set in 1975 for approximately four dollars. Value $45, N.I.B. $100.

Matchbox Truck: This Matchbox truck with its three industrial yellow tractors was not available as a set from Lesney, but they could be added to your collection if you had an extra truck and three industrial tractors that you weren't using. Value $45, N.I.B. n/a.

Matchbox Big Mx Ford Factory: If you like toys out of the ordinary, how about these little Ford Matchbox tractors from London, England. This set came with a little factory that had a conveyor belt for the new tractors to ride through and a Ford tractor-trailer unit parked at the other end to haul the new machines to Ford dealerships throughout England. If this wasn't unusual enough for a toy, try to explain why the tractors were painted orange instead of the traditional yellow and blue. This set was made in 1972, and seems to be one of those hard to find Ford toys here in the United States. Value $150, N.I.B. $300.

1960 Mack Truck with Ford Tractors: In 1998, Ford toy collectors found this interesting combination of tractors loaded on the bed of a 1960 Mack Truck. The load included a Ford 9N, a Ford 901, and an English Fordson Major. Ertl had this set manufactured in China and stores sold them in the United States for twenty dollars. Value $12, N.I.B. $20.

1960 Mack with 1962 Ford 4000s: In 1998, Ertl made a nice set of Ford tractors on the bed of a 1960 Mack truck-trailer unit. Two of the tractors were pre-1960 models and in my opinion, looked a little strange to be on a dealer Ford truck of the 1960 era. Because of that, I decided to combine what I thought to be a satisfying combination. The set consists of the 1960 Mack tractor-trailer unit with a load of four 1962 industrial yellow 4000 Ford tractors. This set was combined at the cost of twenty-five dollars. Value $25, N.I.B. n/a.

New Holland dealership pickup with 8N: This little New Holland dealership pickup with trailer and 8N Ford was available as a set in 1996 for approximately eleven dollars. It was manufactured by Ertl in the 1/64 scale. I don't think that they were made with the Ford New Holland dealership pickup or with 9N tractors, but that could have easily happened. Value $11, N.I.B. $15.

1960 Ford Dealer Pickup: A more popular bank produced in late 1994 by Ertl for New Holland dealers was the 1960 Ford Dealer's Pickup in the 1/20 scale. It sold for eighteen dollars and quickly disappeared from dealer's shelves. Value $25, N.I.B. $35.

Equipment Hauling Set: In 1992, Ertl took their 1/64 scale Ford loader-backhoe and teamed it up with a Ford Louisville 9000 tractor-trailer unit to make a nice construction hauling set. This set sold for eighteen dollars. Value $18, N.I.B. $35.

Nostalig™ Golden Jubilee: This little 1953 Golden Jubilee may be classified as rare in some literature; that is, if you can find it listed at all. At any rate, the available information is rare, but I'll share what I know about this Ford tractor. I purchased this tractor from a miniature toy mail order outlet in 1985. It wasn't even listed in their catalog, but a phone call assured me they could get the tractor. This Ford model was made in 1984 and possibly only that year. The realistic engine, three point hitch, and gearshift made this a very detailed toy. This toy was more like a model than a toy due to its fragility. Value $50, N.I.B. $125.

The F6 Ford truck the Jubilee is sitting on was made by First Gear of Peosta, Iowa. The truck has a custom built flatbed and tractor decals were applied to the side. All the materials that customized the truck cost twenty dollars. Value $25, N.I.B. n/a.

1953 Ford Dealer Trucks: I won't even try to describe these two trucks, except to say I can sit and look at them for hours. They were made by First Gear of Peosta, Iowa. The truck on the left is a model of a 1953 Ford, F100 pickup with Ford tractor dealer decals on the doors. The big truck to the right shows the 1953 Golden Jubilee graphics. The letter "C" in the Ford C-600 stood for a cab over design. The driver's compartment was partly built over the engine, this to allow for a shorter overall length of the truck. Both trucks were made in 1995 at the cost of thirty dollars each. Value $30, N.I.B. $45 each.

Equipment Hauling Set with 9700 Tractors: In 1981, Ertl made this 1/64 scale equipment hauling set with two 9700 tractors. It was a nice set with a Ford CL 9000 tractor-trailer unit that was manufactured in Hong Kong. This set sold at Ford Tractor dealerships for seven dollars. Value $45, N.I.B. $75.

Equipment Hauling Set with TW20 Tractors: In 1982, Ertl made this 1/64 scale equipment hauling set with two TW20 tractors. It was a nice set with a Ford CL 9000 tractor-trailer unit that was manufactured in Hong Kong. This set sold at Ford tractor dealerships for eight dollars. Value $45, N.I.B. $75.

Ford TW35 Equipment Hauling Set: The 1/64 scale toys seen in the photograph were all made by Ertl, but not offered in a set as shown here. The early Ford tractor-trailer unit is shown here with its load of two blue decal TW35 Ford farm tractors which were from the 1990s. Value $50, N.I.B. n/a.

Equipment Hauling Set with TW35 Tractors: In 1989 Ertl made this 1/64 scale equipment hauling set with two black decal TW35 tractors. It was a nice set with a Ford LTL 9000 tractor-trailer that was available at Ford dealerships for eleven dollars. Value $35, N.I.P. $45.

Equipment Hauling Set With 8730 Tractors: In 1991, Ertl made this 1/64 scale equipment hauling set that sold for eighteen dollars. Value $20, N.I.P. $30.

Ford 8240 Equipment Hauling Set The 1/64 scale toys shown in the photograph were all made by Ertl, but not offered in a set. The Ford tractor-trailer unit is shown with its load of two 8240 Ford farm tractors from the 1990s. This combined set could be purchased for eighteen dollars. Value $20, N.I.P. n/a.

Equipment Hauling Set with Genesis 8970 Tractors: This set was manufactured by Ertl in 1995 in the 1/64 scale. This set sold at Ford tractor dealerships for twenty-two dollars. Value $25, N.I.P. $35.

These 1/64 scale toys were made by Ertl in the early 1990s, unfortunately they were never sold in a set as shown in the photograph. The unusual item in this set is the sickle mower connected to the Ford tractor. The mower came in a set with three other small plastic toys, which are seldom seen today. Value $25. N.I.B. n/a.

Ford Dealer Tilt Bed Truck: In 1983 I bought one the these 1/16 scale Ford dealer tilt bed trucks, and as you know, most Ford toys were made in the 1/12 scale. The only 1/16 scale Ford I had at the time was the Fordson "F" and in my opinion, it just didn't look good on this truck. About a week later the truck ended up at the grandparents' house as a toy for my nieces and nephews. Time has a way of making you see things differently. After seventeen years of wondering whether or not I made the right decision, I now own another Ford tilt bed truck. I'm not going to say what this lesson cost me. Value $75, N.I.B. $125.

Ertl Truck with Three Hubley Blades: Shown in the photograph is an Ertl GMC truck with its load of three Hubley blades, no doubt heading to a mid-west Ford dealership Value $750, N.I.B. n/a.

Parts Pay Off 1915 Ford: This truck was manufactured by Ertl around 1992 in the 1/43 scale. The purchase price of this toy was five dollars. Value $10, N.I.B. $15.

Parts Pay Off 1932 Ford: Ertl also made another truck in 1992 in the 1/43 scale. It could be purchased for five dollars. Value $10, N.I.B. $15.

New Holland Geotech 32 Ford Bank: Fiat of Italy began the acquisition of Ford New Holland in the early 1990s. The new company was then given the name New Holland Geotech. Ertl produced the 1932 Ford bank in 1992 to help promote the new company name. A short time later, New Holland Geotech changed its name to New Holland of North America. The average price for this bank was fifteen dollars. Value $20, N.I.B. $25.

New Holland Baler 50 Anniversary Truck: Twenty dollars could purchase this 1/64 scale Ertl truck in 1990. Value $25, N.I.B. $50.

Ford Tractor Parts Van: This 1/64 scale truck was manufactured by Ertl in 1980. It could be purchased for five dollars. Value $25, N.I.B. $45.

Fordson Van: Corgi made this 1/43 scale truck around the year 1970. It sold for two dollars. Value $20, N.I.B. $35.

Fordson Delivery Van: The Model "T" van was manufactured by Ertl in the 1/32 scale. It could be purchased in 1993 for twelve dollars. Value $8, N.I.B. $15.

Ertl 1/64 scale Ford dealership set: I didn't know about this Ertl set when it first came out in 1980. A few years' later, dealers had them priced for twenty five dollars. I thought this price was too cheap so I decided to wait until 1990 when they had gone up in price to fifty dollars before buying one. It is a nice set with a 1/64 scale 9700 tractor, baler, wagon, Ford dealer van, and a Ford Mustang Mach I. All this was centered on a snap together cardboard Ford tractor dealership with a garage and highway. Value $25, N.I.B. $100.

Bibliography

Crilly, Raymond E. and Charles E Burkholder. *International Directory of Model Farm Tractors.* West Chester, Pennsylvania: Schiffer Publishing, Ltd., 1995.

Nolt, Dave. "Zeke Tractor Plays Role in Gift to Former President Bush," *Toy Farmer,* June, 1993.

Pripps, Robert N. *Farm Tractor Color History, Fordson Tractors.* Osceola, Wisconsin: Motorbooks International Publishers and Wholesalers, 1995.

Pripps, Robert N. *Ford Tractors N Series, Fordson, Ford, and Ferguson 1914-1954.* Osceola, Wisconsin: Motorbooks International Publishers and Wholesalers, 1991.

Pripps, Robert N. and Andrew Morland. *Ford and Fordson Tractors.* Osceola, Wisconsin: Motorbooks International Publishers and Wholesalers, 1995.

Toy Tractor Times, Osage, Iowa: August, 1998.

Williams, Michael. *Ford and Fordson Tractors.* Sussex, England: Blandford Press, Ltd., 1985.

Catalogs

Ertl, U.S.A. *Ertl Replicas,* 1983, 1984.
Ertl, U.S.A. *Ford New Holland Catalog,* 1988, 1992.
Scale Models, U.S.A. *Ford New Holland Comin' on Strong,* 1990.
Ertl, U.S.A. Ford New Holland Toys, 1992-1993.
Scale Models, U.S.A. *New Holland Fiftieth Anniversary Catalog,* 1995.
Ertl, U.S.A. *Perfect Match Replica,* 1985, 1987.

Index

A
A.C. Williams, 8
Arcade, 7, 23

B
Bing, 8
Britains, 16, 17, 51, 58, 59, 60, 62, 65, 66, 67, 72, 78, 89, 92, 102, 103
Brooks, 11
Buddy L, 92

C
Corgi, 17, 18, 59, 62, 141
Cragstan, 99, 100, 101
Crescent, 18
Custom, 6, 21, 38, 70, 82, 101, 116

D
Danbury Mint, 15, 33, 51

E
Ertl, 6, 11, 12, 13, 15, 16, 19, 20, 21, 22, 24, 25, 26, 30, 31, 32, 33, 35, 36, 39, 40, 42, 45, 46, 47, 48, 49, 55, 56, 57, 58, 60, 61, 68, 69, 70, 71, 73, 74, 75, 76, 78, 79, 81, 82, 83, 84, 85, 86, 87, 88, 89, 90, 91, 92, 94, 95, 96, 97, 98, 101, 103, 104, 105, 106, 107, 108, 109, 112, 113, 117, 118, 119, 120, 121, 122, 123, 127, 128, 129, 131, 132, 133, 134, 135, 136, 137, 138, 139, 140, 141, 142

F
First Gear, 134
Franklin Mint, 36
Fun Ho, 19

G
Gabriel, 64
Galanite, 61
Gray, 24

H
Hover, 126, 127
Hubley, 25, 43, 44, 45, 52, 53, 54, 63, 64, 115, 124, 125, 126, 128

I
Irwin, 42

J
Jouef, 81

K
Kansas Toy, 11
Kenton, 8

L
Lesney, 19, 130, 131
Lonestar, 71, 72

M
Maisto, 66, 67
Majorette, 61
Marusano, 29
Micro Machines, 10
Modern Toys, 10

N
New Ray, 77
Nostalig, 133

O
Olti, 65

P
Pacesetter, 92, 93
Pioneer, 24
Plasto, 93
Polistil, 76
Popular Imports, 32
Processed Plastics, 75
Product Miniature, 27, 28, 29, 34

R
Riecke, 9

S
Scale Models, 14, 26, 33, 42, 50, 54, 65, 77, 91, 102, 109, 110, 111, 113, 117, 120
Sharp, 58
Siku, 91
Slik, 38, 41
Spec Cast, 79, 80

T
Tenko Bogg, 9
Tootsietoy, 30
Tru-Scale, 35, 45, 46